Biomass for Sustainability: Resource, Technology Conversion and Energy Management

Biomass for Sustainability: Resource, Technology Conversion and Energy Management

Editors

Luis M. López-Ochoa
José P. Paredes-Sánchez

MDPI • Basel • Beijing • Wuhan • Barcelona • Belgrade • Manchester • Tokyo • Cluj • Tianjin

Editors
Luis M. López-Ochoa
University of La Rioja
Spain

José P. Paredes-Sánchez
University of Oviedo
Spain

Editorial Office
MDPI
St. Alban-Anlage 66
4052 Basel, Switzerland

This is a reprint of articles from the Special Issue published online in the open access journal *Sustainability* (ISSN 2071-1050) (available at: https://www.mdpi.com/journal/sustainability/special_issues/BiomassSustain_ResTechnConEnergManag).

For citation purposes, cite each article independently as indicated on the article page online and as indicated below:

LastName, A.A.; LastName, B.B.; LastName, C.C. Article Title. *Journal Name* **Year**, *Volume Number*, Page Range.

ISBN 978-3-0365-4343-7 (Hbk)
ISBN 978-3-0365-4344-4 (PDF)

Contents

sustainability

MDPI

Article

Energy Utilization of Algae Biomass Waste *Enteromorpha* Resulting in Green Tide in China: Pyrolysis Kinetic Parameters Estimation Based on Shuffled Complex Evolution

Lingna Zhong [1,*], Juan Zhang [2] and Yanming Ding [2,3]

1 School of Politics and International Studies, Central China Normal University, Wuhan 430079, China
2 Faculty of Engineering, China University of Geosciences, Wuhan 430074, China; zhangjuan@cug.edu.cn (J.Z.); dingym@cug.edu.cn (Y.D.)
3 State Key Laboratory of Coal Resources and Safe Mining, China University of Mining and Technology, Xuzhou 221116, China
* Correspondence: zhonglingna@mail.ccnu.edu.cn; Tel.: +86-0276-786-8931

Received: 24 February 2020; Accepted: 6 March 2020; Published: 8 March 2020

Abstract: *Enteromorpha* is a species of algae biomass that is spread widely and has resulted in green tides in China in recent years. It was urgent to explore an appropriate method for taking advantage of the ocean waste as an energy supply in the current sustainable development. Pyrolysis, as the first step of thermochemical conversion in energy utilization, was given attention in order to study its behavior based on thermogravimetric experiments over a wide heating-rate range from 5 to 60 K/min. The whole pyrolysis process was divided into three stages: water evaporation, the main components decomposition, and carbonate decomposition. To estimate the detailed kinetic parameters (activation energy, the pre-exponential factor, and reaction order etc.), the Kissinger method was used to establish the original kinetic parameters at different stages and provide the parameter search range for the next heuristic algorithm, and then the Shuffled Complex Evolution optimization algorithm was coupled and first applied to the algae biomass pyrolysis. Eventually, the predicted results of mass loss rate based on the optimized kinetic parameters agreed well with the thermogravimetric experimental data, with the R^2 value being up to 0.92 for all the heating rates.

Keywords: green tide; parameter optimization; pyrolysis kinetics; Shuffled Complex Evolution; Kissinger method

1. Introduction

Many countries are surrounded by wide coastal areas and territorial seas. There are a lot of algae biomass species living along 32 thousand kilometers of coastline in China. With a short life cycle and fast breeding, they are easy to breed [1,2]. However, due to global climate change and water body eutrophication, the accumulating of massive amounts of *Enteromorpha* occurs, named "green tide", which has become one of the major marine environmental problems all over the world, resulting in the damage of marine ecosystems, waterways, biodiversity, and even the coastal aquaculture, fishery, and tourism industry [3]. For example, green tide broke out in Qingdao (China's yellow sea) in the summer of the 2008 Beijing Olympics, and more than 60 million tons of two green algae species (*Enteromorpha clathrata* and *Enteromorpha compressa*) grew very rapidly [4]. Thirty percent of the sailing competition area was covered by *Enteromorpha* with a direct economic loss of 1.322 billion RMB [3]. Such green tide is not accidental and it continues to thrive today. Green tide appears frequently along coastlines and has caused extensive ocean pollution and big economic losses over the years [3]. Therefore, it is very urgent to explore how to solve this problem.

Energy supply is one of the greatest challenges for the sustainable development of the current society [5]. The exploration and development of safe and sustainable alternatives to fossil fuels are given important global priorities [6]. Accordingly, renewable energy has gained more attention to address this problem. Algae biomass which lives in the largest aquatic environment area of the earth, as an important kind of renewable energy, shows several main advantages compared with terrestrial biomass, including no competition for arable land, cultivation in fresh or saltwater, no need for fertilizer or pesticide application, year-round production capability, high photosynthetic capacity, and high CO_2 capture efficiency. Moreover, it also reveals other benefits, such as sequestering environmental contaminants [7,8]. Therefore, algae biomass, categorized as a third-generation biofuel resource, is not only ocean waste but also one of renewable energy [3,9]. Furthermore, it has become the biomass resource with most potential and is currently cultivated on an industrial scale for energy utilization [10]. Namely, a practical and feasible technology to solve the problem of green tide is to take advantage of these algae wastes for energy supply. At present, thermal-chemical conversion processes are regarded as an important energy utilization approach to algal biomass, especially pyrolysis [11–14]. Pyrolysis is not only a direct pathway for biomass thermal conversion, but also the first step in combustion and gasification reactions [7,15–18]. Thus, an improved understanding of this process would be helpful for thermochemical system design. Following this, an evaluation of the pyrolysis behaviors and kinetic characteristics of algae biomasses is needed.

Thermogravimetric analysis (TGA) is a technology widely used to explore the thermal conversion of biomass pyrolysis as a function of temperature and/or time in a controlled atmosphere [7,12], which will be applied here. Moreover, various reactions might be involved in the pyrolysis process with lots of kinetic parameters (activation energy, pre-exponential factor, reaction order etc.) need to be calculated. It is difficult for traditional kinetic methods to estimate so many parameters, therefore heuristic algorithms have to be considered to solve this problem. The Shuffled Complex Evolution (SCE) optimization algorithm, as a classic heuristic algorithm, is proposed by Duan et al. [19,20]. Its applicability and effectiveness to terrestrial lignocellulosic biomass pyrolysis have been verified in our previous study [21,22]. However, the optimization algorithm has not been applied to algae biomass pyrolysis. In order to fill the gaps in knowledge, *Enteromorpha clathrate*, as a typical algae biomass in green tide, is used in our current study. Thermogravimetric analysis is conducted at various heating rates to explore its pyrolysis behaviors, and its kinetic parameters are estimated by coupling the Kissinger method and the SCE optimization algorithm.

2. Materials and Methods

2.1. Thermogravimetric Measurements

Enteromorpha clathrate was collected from the Qingdao coast and sun dried for 10 days. The detailed properties of this sample, including ultimate analysis and biochemical composition are listed in Table 1. To conduct thermogravimetric experiments, all the samples were milled to powder with a size of less than 0.01 mm. A SDT Q600 thermal analyzer (TA Instruments, New Castle, DE, USA) was applied in the pyrolysis process with five heating rates (5–60 K/min) from 300 to 1200 K. All the samples were evenly placed in an Alumina cup without a lid, and a high purity nitrogen flow (100 mL/min) was used for all the experimental runs.

Table 1. Properties of *Enteromorpha clathrate* [1] (% mass, dry basis).

Ultimate Analysis	Value	Biochemical Composition	Value
C	22.74	Lipids	1.28
H	6.27	Proteins	23.99
N	3.14	Carbohydrates	40.00
S	1.27		
O	7.89		

[1] Information from Wang et al. [4].

2.2. Pyrolysis Kinetics

A pyrolysis kinetic equation can be expressed based on conversion rate α as follows:

$$\frac{d\alpha}{dt} = k(T)f(\alpha) \tag{1}$$

where $k(T)$ and $f(\alpha)$ are the reaction rate constant and reaction mechanism function, respectively. Whereby, α and $k(T)$ can be further calculated via the following equations:

$$\alpha = \frac{m_0 - m_t}{m_0 - m_\infty} \tag{2}$$

$$k(T) = A\exp\left(\frac{-E}{RT}\right) \tag{3}$$

where m_0, m_t, and m_∞ represent the sample mass at the initial time, time t, and the end, respectively. A and E are the pre-exponential factor and activation energy of the reaction, respectively.

Then Equation (1) can be written as

$$\frac{d\alpha}{dT} = \frac{A}{\beta}f(\alpha)\exp\left(\frac{-E_a}{RT}\right) \tag{4}$$

where the linear heating rate $\beta = dT/dt$.

Based on the basic pyrolysis kinetic equation, the Kissinger method was applied to calculate the activation energy and pre-exponential factor at the peak locations, as a reference of the parameter search range of the SCE algorithm.

2.2.1. Kissinger Method

Assuming that the reaction-order model should be responsible for our current reaction, the reaction mechanism function was defined as $f(\alpha) = (1 - \alpha)^n$, and then the time derivative of Equation (4) at peak locations could be expressed as

$$\frac{E}{RT_p^2}\frac{dT}{dt} = An\left(1 - \alpha_p\right)^{n-1}\exp\left(\frac{-E}{RT_p}\right) \tag{5}$$

Kissinger regarded that $n(1 - \alpha_p)^{n-1}$ was independent of the heating rate β, with the value being approximately equal to 1. Then the Kissinger kinetic equation was obtained by taking the logarithm of the following equation:

$$\ln\left(\frac{\beta}{T_p^2}\right) = \ln\frac{AR}{E} - \frac{E}{R}\frac{1}{T_p} \tag{6}$$

The plot of $ln(\beta/T^2_p)$ versus $1/T$ gave a straight line, whose slope and intercept could be used to determine the activation energy E and pre-exponential factor A.

2.2.2. SCE Optimization Algorithm

The SCE global optimization algorithm is an effective method for optimizing the pyrolysis kinetic parameters of biomass pyrolysis. This has been proved in our previous study [21–23]. There are four concepts for this algorithm: the combination of probabilistic and deterministic approaches, clustering, the systematic evolution of a complex of points spanning the space in the direction of global improvement, and competitive evolution [19]. In the optimization process, an initial population composed of parameters that need to be optimized is generated. Next, these parameters (called individuals) are ranked and partitioned into multiple complexes. Then each complex is evolved independently according to the Competitive Complex Evolution algorithm and the second generation is produced by combining all the individuals in each complex. For each generation, the fittest individuals can be survived based on the objective function until convergence is achieved. The more detailed optimization process can be referred to Duan and Chaos et al. [19,20,24]. For the current thermogravimetric experiment, the objective function φ_{mlr} was the least error of mass loss rate (MLR) between the predicted results and experimental data:

$$\varphi_{mlr} = \sum_{j=1}^{N} \left[\frac{\sum\limits_{k=1}^{\lambda} \left(MLR_{\text{pred,k}} - MLR_{\text{exp,k}} \right)^2}{\sum\limits_{k=1}^{\lambda} \left(MLR_{\text{exp,k}} - \frac{1}{\lambda} \sum\limits_{p=1}^{\lambda} MLR_{\text{exp,p}} \right)^2} \right] \tag{7}$$

where N and λ are the number of experiments and experimental data points for each experiment, respectively.

3. Results and Discussion

3.1. Thermogravimetric Analysis

The thermogravimetric experimental data, including the conversion rate and mass loss rate of *Enteromorpha clathrate* pyrolysis at various heating rates, are shown in Figure 1. Ross et al. [25] found that there were significant differences in the pyrolysis properties between algae biomass and terrestrial lignocellulosic biomass. Different from lignocellulosic biomass, whose main components are hemicellulose, cellulose, and lignin [26,27], the main components of algal biomass are protein, carbohydrate, and lipid.

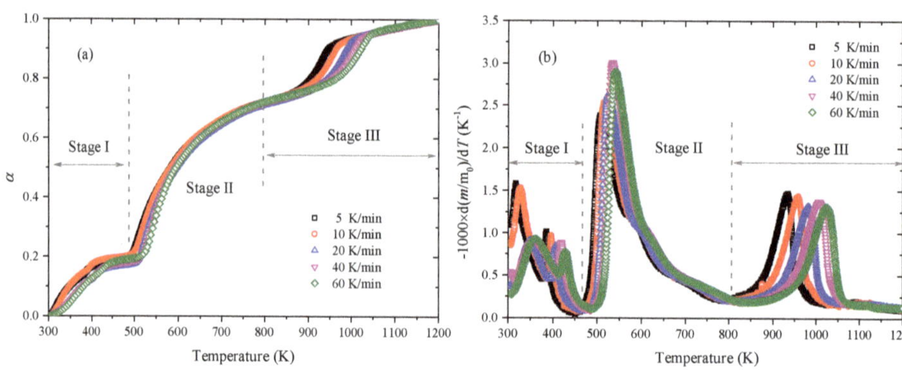

Figure 1. Pyrolysis behaviors at different heating rates: (**a**) conversion rate and (**b**) mass loss rate.

Figure 1 shows that the main pyrolysis process was divided into three parts: Stage I in the temperature range of 300–460 K, Stage II in the temperature range of 460–800 K, and Stage III in the temperature range of 800–1100 K, which is marked in the figure. In Stage I, a dehydration process—namely, the evaporation of water happens—which was also observed by Wang, Kim and Zhao et al. [4,28,29] in a similar temperature range. The total conversion rate in this stage was about 20%. In Stage II, the main components of *Enteromorpha clathrate* started to decompose. Namely, protein, carbohydrate, and lipid should have been responsible for the pyrolysis in this stage with about a 50% total conversion rate. The obvious peaks were observed within the temperature range of 520–570 K. The higher the heating rate was, the greater the mass loss rate of peak was. In Stage III, the reaction occurred at a high temperature, which might be attributed to the carbonate decomposition or volatile metal loss [4]. The total conversion rate in this stage was about 30% and the peak location was between 930 and 1030 K. Differently from the change trend of the peak in Stage II, the higher the heating rate was, the smaller the mass lose rate of peak was in Stage III.

In the whole pyrolysis process, the main decomposition reactions of *Enteromorpha clathrate* happened in Stage II and III, which can be expressed as

$$\begin{aligned} Component_A &\rightarrow v_A residue + (1 - v_A) volatiles \\ Component_B &\rightarrow v_B residue + (1 - v_B) volatiles \end{aligned} \tag{8}$$

where v is the residue yield, $Component_A$ and $Component_B$ are the hypothetical reactants that should have been responsible for Stages II and III, respectively.

3.2. Kinetic Analysis Based on the Kissinger Method

To calculate the activation energy and pre-exponent factor, the peak location should first be established, then the second derivative of m/m_0 (DTG) is used here [23,30,31]:

$$|\text{DDTG}| = \left| \frac{d^2(m/m_0)}{dT^2} \right| = \left| \sum_{i=1}^{3} Y_i \frac{d^2\alpha_i}{dT^2} \right| \geq 0 \tag{9}$$

Li et al. [30] stressed that if a major component was near the maximum decomposition rate, then the value of |DDTG| would be likely to drop rapidly to a local minimum (LM). The curve of |DDTG| in Stages II and III at 40 and 60 K/min is shown in Figure 2 and the local minimum locations (LM) are still marked. By coupling the DTG and |DDTG| curves, it was convenient to establish the peak temperature, namely 533.97 and 1006.57 K at 40 K/min, and 540.79 and 1023.11 K at 60 K/min. The detailed peak temperature for all the heating rates is listed in Table 2.

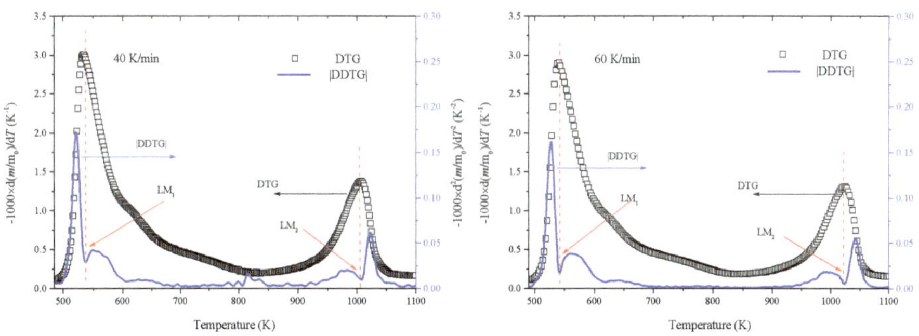

Figure 2. Estimation of peak locations based on second derivative of m/m_0 (|DDTG|).

Table 2. Peak temperatures at different heating rate.

Temperature (K)	Heating Rate (K/min)				
	5	10	20	40	60
T_{p1}	507.59	517.46	525.09	533.97	540.79
T_{p2}	935.46	957.10	980.27	1006.57	1023.11

Based on the established peak locations and Equation (6), the slope and intercept were obtained in Figure 3, and the activation energy and pre-exponent factor were calculated by the Kissinger method: E_{p1} = 166.45 kJ/mol, E_{p2} = 209.34 kJ/mol, $A_{p1} = 8.50 \times 10^{11}$ s^{-1}, and $A_{p2} = 1.21 \times 10^{6}$ s^{-1}, respectively.

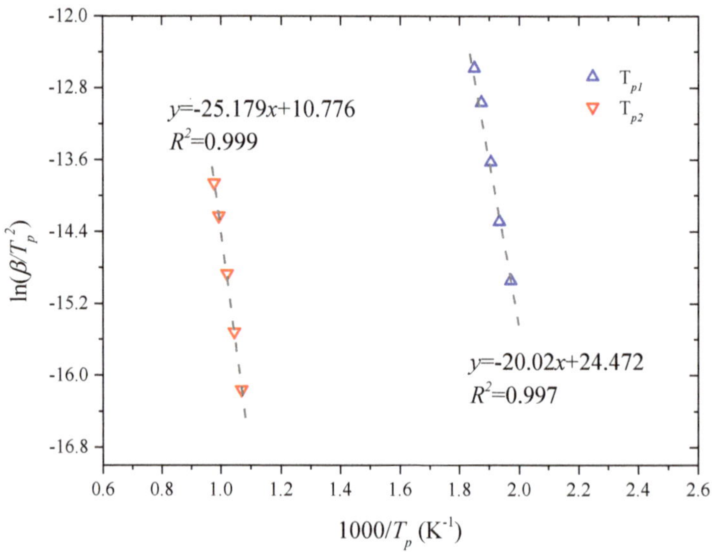

Figure 3. Kissinger plots for peak locations.

3.3. Estimation of Kinetic Parameters by SCE

The obtained activation energy and pre-factor were defined as the original value V_0, and the parameter search range of the SCE was set to 0.5–1.5V_0, as listed in Table 3.

Table 3. Optimized parameters by Shuffled Complex Evolution at multiple heating rates.

Parameters	Initial Values	Search Range	Optimized Values
$Y_{A,0}$	0.50	[0.25, 0.75]	0.38
$\ln(A_A/s^{-1})$	27.47	[13.73, 41.20]	41.19
E_A (kJ/mol)	166.45	[83.22, 249.67]	199.85
n_A	1.00	[0.00, 10.00]	9.14
v_A	0.50	[0.00, 1.00]	0.19
$Y_{B,0}$	1- $Y_{A,0}$	[0.00, 1.00]	0.62
$\ln(A_B/s^{-1})$	14.00	[7.00, 21.00]	20.95
E_B (kJ/mol)	209.34	[104.67, 314.01]	208.91
n_B	1.00	[0.00, 10.00]	0.66
v_B	0.50	[0.00, 1.00]	0.85

For Equation (8), the reaction rates of the hypothetical reactants and produced residue could be represented by the nth order reaction-order model

$$\frac{dY_i}{dt} = -Y_{i,0}\left(\frac{Y_i}{Y_{i,0}}\right)^{n_i} A_i \exp\left(-\frac{E_i}{RT}\right)(i = 1,2) \tag{10}$$

$$\frac{dY_{residue}}{dt} = -\sum_{i=1}^{2} v_i \frac{dY_i}{dt} \tag{11}$$

Then the predicted value of total mass loss rate was expressed as

$$MLR = \frac{d(m/m_0)}{dT} = \frac{1}{\beta}\left(\sum_{i=1}^{2} \frac{dY_i}{dt} + \frac{dY_{residue}}{dt}\right) \tag{12}$$

The above predicted results were compared with thermogravimetric experimental data by the object function of Equation (7). There were 9 kinetic parameters that needed to be optimized at the same time. Eventually, the optimized results were obtained by SCE based on all five heating rates, as shown in Figure 4, taking the results of 5–40 K/min, for example. The mass loss rates of separate component are further displayed in Figure 5, taking the results of 5 and 10 K/min, for example. Moreover, the optimized parameters are listed in Table 3.

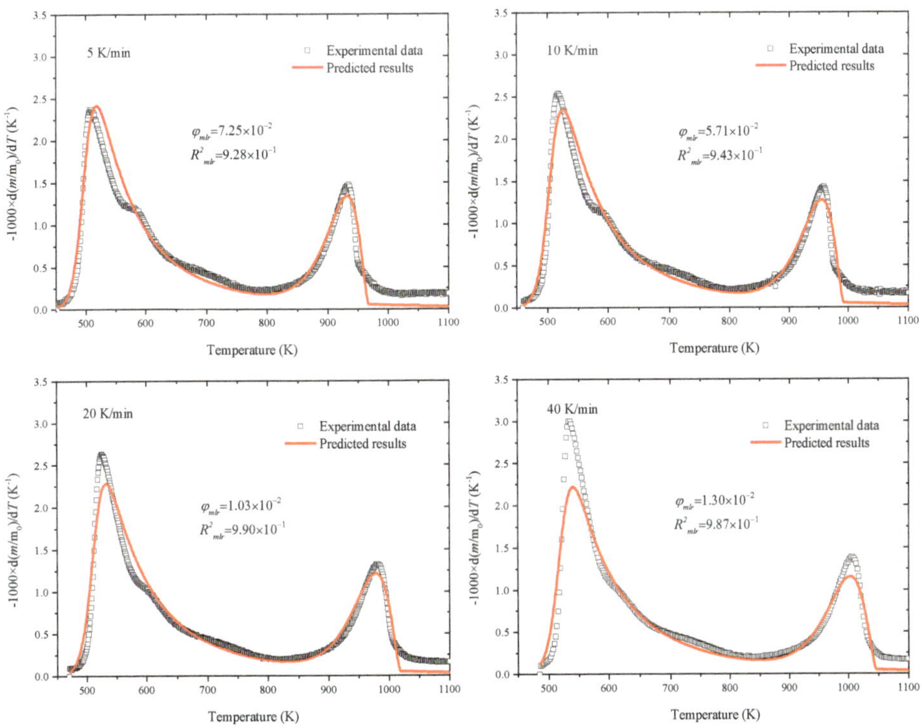

Figure 4. Predicted mass loss rates based on the optimized parameters (lines) compared with experimental data (symbols).

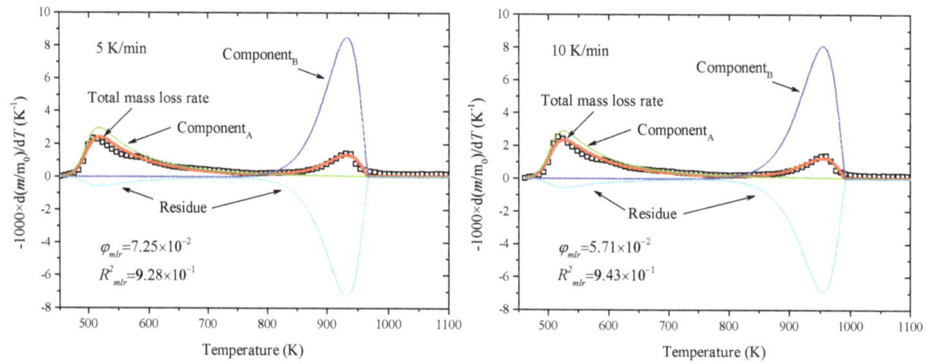

Figure 5. Predicted mass loss rates of separated components.

This shows that the predicted results agreed well with the experiment data with R^2 being up to 0.92, especially the peak locations, which were captured exactly in both Stages II and III. The activation energy and pre-exponent factor in Stages II and III were E_{p1} = 199.85 kJ/mol, A_{p1} = 7.75 × 10^{17} s^{-1}, E_{p2} = 208.91 kJ/mol, and A_{p2} = 1.25 × 10^9 s^{-1}, respectively. The optimized parameters could be used in the pyrolysis reactor design and numerical simulation. Our previous study proved the applicability of optimized parameters in the pyrolysis model coupled with mass and heat transfer [32] and even the complicated direct combustion process in the field of numerical simulation [33].

4. Conclusions

A series of thermogravimetric experiments from 5 to 60 K/min was conducted to explore the pyrolysis behaviors of the typical algae biomass *Enteromorpha clathrate*, which results in green tide in China. The whole pyrolysis behavior was composed of three stages, including water evaporation, main-component (protein, carbohydrate, and lipid) decomposition and carbonate decomposition. To explore the kinetic parameters (such as the activation energy, pre-exponential factor, and reaction order etc.) in the main reaction stages, the second derivative of mass loss rate was used to establish the peak locations. Then a traditional kinetic method called Kissinger was used to calculate the original kinetic parameters according to the above peak locations. Furthermore, the Shuffled Complex Evolution algorithm was coupled and first applied to the algae biomass pyrolysis. Eventually, the optimized kinetic parameters were obtained based on all the heating rates. The predicted mass loss rates agreed well with the thermogravimetric experimental data, with R^2 being up to 0.92.

Author Contributions: Conceptualization, L.Z. and Y.D.; methodology, L.Z. and Y.D.; validation, L.Z., J.Z., and Y.D.; investigation, L.Z.; data curation, J.Z.; writing—original draft preparation, L.Z.; writing—review and editing, J.Z. and Y.D.; supervision, L.Z. All authors have read and agreed to the published version of the manuscript.

Funding: This research was funded by the National Natural Science Foundation of China, grant numbers 51806202 and 51806106, the Natural Science Foundation of Hubei Province of China, grant number 2018CFB352 and the Research Fund of The State Key Laboratory of Coal Resources and Safe Mining, CUMT, grand number SKLCRSM19KF021.

Conflicts of Interest: The authors declare no conflict of interest.

References

1. Wang, S.; Jiang, X.; Wang, N.; Yu, L.; Li, Z.; He, P. Research on pyrolysis characteristics of seaweed. *Energy Fuels* **2007**, *21*, 3723–3729. [CrossRef]
2. Zhao, B.; Fang, Y.; Wu, K.; Zhang, F.; Wang, J. A Method of Large-Scale Resource Utilization of Algae—Eutrophic Waste from Lake Chao, China: Preparation and Performance Optimization of Composite Packaging Materials. *Sustainability* **2019**, *11*, 6462. [CrossRef]

3. Gao, X.; Xing, W.; Zhou, J.; Wang, G.; Zhuo, S.; Liu, Z.; Xue, Q.; Yan, Z. Superior capacitive performance of active carbons derived from *Enteromorpha prolifera*. *Electrochim. Acta* **2014**, *133*, 459–466. [CrossRef]
4. Wang, S.; Jiang, X.; Han, X.; Liu, J. Combustion characteristics of seaweed biomass. 1. Combustion characteristics of *Enteromorpha clathrata* and *Sargassum natans*. *Energy Fuels* **2009**, *23*, 5173–5178. [CrossRef]
5. Li, D.; Chen, L.; Zhao, J.; Zhang, X.; Wang, Q.; Wang, H.; Ye, N. Evaluation of the pyrolytic and kinetic characteristics of *Enteromorpha prolifera* as a source of renewable bio-fuel from the Yellow Sea of China. *Chem. Eng. Res. Des.* **2010**, *88*, 647–652. [CrossRef]
6. Hua, M.-Y.; Li, B.-X. Co-pyrolysis characteristics of the sugarcane bagasse and *Enteromorpha prolifera*. *Energy Convers. Manag.* **2016**, *120*, 238–246. [CrossRef]
7. Ceylan, S.; Goldfarb, J.L. Green tide to green fuels: TG–FTIR analysis and kinetic study of Ulva prolifera pyrolysis. *Energy Convers. Manag.* **2015**, *101*, 263–270. [CrossRef]
8. Guo, Y.; Tan, C.; Wang, P.; Sun, J.; Li, W.; Zhao, C.; Lu, P. Structure-performance relationships of magnesium-based CO_2 adsorbents prepared with different methods. *Chem. Eng. J.* **2020**, *379*, 122277. [CrossRef]
9. Pal, P.; Chew, K.W.; Yen, H.-W.; Lim, J.W.; Lam, M.K.; Show, P.L. Cultivation of Oily Microalgae for the Production of Third-Generation Biofuels. *Sustainability* **2019**, *11*, 5424. [CrossRef]
10. Wang, S.; Jiang, D.; Cao, B.; Qian, L.; Hu, Y.; Liu, L.; Yuan, C.; Abomohra, A.E.-F.; He, Z.; Wang, Q. Bio-char and bio-oil characteristics produced from the interaction of *Enteromorpha clathrate* volatiles and rice husk bio-char during co-pyrolysis in a sectional pyrolysis furnace: A complementary study. *J. Anal. Appl. Pyrolysis* **2018**, *135*, 219–230. [CrossRef]
11. Wang, S.; Cao, B.; Liu, X.; Xu, L.; Hu, Y.; Afonaa-Mensah, S.; Abomohra, A.E.-F.; He, Z.; Wang, Q.; Xu, S. A comparative study on the quality of bio-oil derived from green macroalga *Enteromorpha clathrata* over metal modified ZSM-5 catalysts. *Bioresour. Technol.* **2018**, *256*, 446–455. [CrossRef] [PubMed]
12. Perrot, J.-F.; Subiantoro, A. Municipal waste management strategy review and waste-to-energy potentials in New Zealand. *Sustainability* **2018**, *10*, 3114. [CrossRef]
13. Xu, Z.; Xiao, X.; Fang, P.; Ye, L.; Huang, J.; Wu, H.; Tang, Z.; Chen, D. Comparison of Combustion and Pyrolysis Behavior of the Peanut Shells in Air and N2: Kinetics, Thermodynamics and Gas Emissions. *Sustainability* **2020**, *12*, 464. [CrossRef]
14. Chen, R.; Li, Q.; Zhang, Y.; Xu, X.; Zhang, D. Pyrolysis kinetics and mechanism of typical industrial non-tyre rubber wastes by peak-differentiating analysis and multi kinetics methods. *Fuel* **2019**, *235*, 1224–1237. [CrossRef]
15. Kong, D.; Peng, R.; Sun, X.; Zhang, J.; Ping, P.; Du, J. Study of the influence of crude oil on the spontaneous combustion risk of sulfurized rust in crude oil tanks. *Fuel* **2019**, *255*, 115816. [CrossRef]
16. Jiang, L.; Zhang, D.; Li, M.; He, J.-J.; Gao, Z.-H.; Zhou, Y.; Sun, J.-H. Pyrolytic behavior of waste extruded polystyrene and rigid polyurethane by multi kinetics methods and Py-GC/MS. *Fuel* **2018**, *222*, 11–20. [CrossRef]
17. Zhou, Y.; Bu, R.; Gong, J.; Yan, W.; Fan, C. Experimental investigation on downward flame spread over rigid polyurethane and extruded polystyrene foams. *Exp. Therm. Fluid Sci.* **2018**, *92*, 346–352. [CrossRef]
18. Kong, B.; Wang, E.; Lu, W.; Li, Z. Application of electromagnetic radiation detection in high-temperature anomalous areas experiencing coalfield fires. *Energy* **2019**, *189*, 116144. [CrossRef]
19. Duan, Q.; Gupta, V.K.; Sorooshian, S. Shuffled complex evolution approach for effective and efficient global minimization. *J. Optim. Theory Appl.* **1993**, *76*, 501–521. [CrossRef]
20. Duan, Q.; Sorooshian, S.; Gupta, V.K. Optimal use of the SCE-UA global optimization method for calibrating watershed models. *J. Hydrol.* **1994**, *158*, 265–284. [CrossRef]
21. Ding, Y.; Wang, C.; Chaos, M.; Chen, R.; Lu, S. Estimation of beech pyrolysis kinetic parameters by shuffled complex evolution. *Bioresour. Technol.* **2016**, *200*, 658–665. [CrossRef] [PubMed]
22. Ding, Y.; Zhang, J.; He, Q.; Huang, B.; Mao, S. The application and validity of various reaction kinetic models on woody biomass pyrolysis. *Energy* **2019**, *179*, 784–791. [CrossRef]
23. Ding, Y.; Huang, B.; Li, K.; Du, W.; Lu, K.; Zhang, Y. Thermal interaction analysis of isolated hemicellulose and cellulose by kinetic parameters during biomass pyrolysis. *Energy* **2020**, *195*, 117010. [CrossRef]
24. Chaos, M. Spectral aspects of bench-scale flammability testing: Application to hardwood pyrolysis. *Fire Saf. Sci.* **2014**, *11*, 165–178. [CrossRef]

25. Ross, A.; Jones, J.; Kubacki, M.; Bridgeman, T. Classification of macroalgae as fuel and its thermochemical behaviour. *Bioresour. Technol.* **2008**, *99*, 6494–6504. [CrossRef]
26. Ding, Y.; Ezekoye, O.A.; Lu, S.; Wang, C.; Zhou, R. Comparative pyrolysis behaviors and reaction mechanisms of hardwood and softwood. *Energy Convers. Manag.* **2017**, *132*, 102–109. [CrossRef]
27. Ding, Y.; Zhang, W.; Yu, L.; Lu, K. The accuracy and efficiency of GA and PSO optimization schemes on estimating reaction kinetic parameters of biomass pyrolysis. *Energy* **2019**, *176*, 582–588. [CrossRef]
28. Kim, S.-S.; Ly, H.V.; Kim, J.; Choi, J.H.; Woo, H.C. Thermogravimetric characteristics and pyrolysis kinetics of Alga Sagarssum sp. biomass. *Bioresour. Technol.* **2013**, *139*, 242–248. [CrossRef]
29. Zhao, H.; Yan, H.; Dong, S.; Zhang, Y.; Sun, B.; Zhang, C.; Ai, Y.; Chen, B.; Liu, Q.; Sui, T. Thermogravimetry study of the pyrolytic characteristics and kinetics of macro-algae *Macrocystis pyrifera* residue. *J. Therm. Anal. Calorim.* **2013**, *111*, 1685–1690. [CrossRef]
30. Li, K.-Y.; Huang, X.; Fleischmann, C.; Rein, G.; Ji, J. Pyrolysis of medium-density fiberboard: Optimized search for kinetics scheme and parameters via a genetic algorithm driven by Kissinger's method. *Energy Fuels* **2014**, *28*, 6130–6139. [CrossRef]
31. Ding, Y.; Zhang, Y.; Zhang, J.; Zhou, R.; Ren, Z.; Guo, H. Kinetic parameters estimation of pinus sylvestris pyrolysis by Kissinger-Kai method coupled with Particle Swarm Optimization and global sensitivity analysis. *Bioresour. Technol.* **2019**, *293*, 122079. [CrossRef] [PubMed]
32. Ding, Y.; Ezekoye, O.A.; Zhang, J.; Wang, C.; Lu, S. The effect of chemical reaction kinetic parameters on the bench-scale pyrolysis of lignocellulosic biomass. *Fuel* **2018**, *232*, 147–153. [CrossRef]
33. Ding, Y.; Fukumoto, K.; Ezekoye, O.A.; Lu, S.; Wang, C.; Li, C. Experimental and numerical simulation of multi-component combustion of typical charring material. *Combust. Flame* **2020**, *211*, 417–429. [CrossRef]

sustainability

MDPI

Article

Bioenergy for a Cleaner Future: A Case Study of Sustainable Biogas Supply Chain in the Malaysian Energy Sector

Nur Izzah Hamna A. Aziz [1], Marlia M. Hanafiah [1,2,*], Shabbir H. Gheewala [3,4] and Haikal Ismail [1,5]

1 Department of Earth Sciences and Environment, Faculty of Science and Technology, Universiti Kebangsaan Malaysia, Bangi 43600, Selangor, Malaysia; izzahhamna@ukm.edu.my (N.I.H.A.A.); haikal.b.ismail@gmail.com (H.I.)
2 Centre for Tropical Climate Change System, Institute of Climate Change, Universiti Kebangsaan Malaysia, Bangi 43600, Selangor, Malaysia
3 The Joint Graduate School of Energy and Environment, Centre of Excellence on Energy Technology and Environment, King Mongkut's University of Technology Thonburi, Bangkok 10140, Thailand; shabbir_g@jgsee.kmutt.ac.th
4 Center of Excellence on Energy Technology and Environment, PERDO, Bangkok 10140, Thailand
5 School of Technology Management and Logistics, College of Business, Universiti Utara Malaysia, Sintok 06010, Kedah, Malaysia
* Correspondence: mhmarlia@ukm.edu.my

Received: 18 March 2020; Accepted: 6 April 2020; Published: 16 April 2020

Abstract: A life cycle assessment (LCA)-based environmental sustainability evaluation conceptual framework of biogas production has been proposed to improve the sustainability of biogas supply chains. The conceptual framework developed in this study can be used as a guideline for the related stakeholders and decision makers to improve the quality and enhance the sustainability of biogas production in Malaysia as well as promoting biogas as a clean, reliable and secure energy. A case study on an LCA analysis of a zero waste discharge treatment process has been conducted. In the zero discharge treatment system, biogas can be produced with a maximum water recycle and reuse. It was indicated that the biogas production and zero discharge treatment of a palm oil mill effluent were environmentally sustainable as the system utilized organic waste to produce bioenergy and achieved zero discharge. However, there were other aspects that should be taken into consideration, particularly regarding the sources of electricity and upstream activity, to ensure the sustainability of the system holistically.

Keywords: clean technology; renewable energy; life cycle assessment; zero discharge; waste treatment; sustainability

1. Introduction

Energy plays a crucial and challenging role in sustainable development. The Malaysian government has been continuously reviewing its energy policy to ensure the sustainability of energy resources due to the increase of energy demand [1,2]. The emerging of renewable energy production to replace fossil fuels consumption could certainly reduce environmental pollution because it provides a low-carbon energy system [3]. Nonetheless, the production of renewable energy consists of input and output flows and operational processes which may influence its performance. It is essential to obtain renewable energy in sustainable ways in order to achieve sustainable development as well as promoting economic growth in the country. Clean energy, sustainable cities and communities, responsible consumption and production and climate action are included in the 17 Sustainable

Development Goals (SDGs). Griggs et al. [4] argued that planetary stability should be integrated in the United Nations' targets, which are to eliminate poverty and protect the earth's life-support system.

Biogas is produced by the indirect conversion of solar energy stored in natural organic matter into a gaseous energy carrier by anaerobic fermentation; therefore, biogas is the end product of microbiological fermentation [5,6]. The main combustible component in biogas is methane (CH_4), and it also contains significant amounts of carbon dioxide (CO_2) and other trace gases [7,8]. Biogas can be used for heating, electricity generation, as a fuel for vehicle or distributed via the natural gas grid [9,10]. Furthermore, the digestate produced by the anaerobic digestion process can be used as a biofertilizer in agriculture, thus reducing the need for chemical fertilizers [11,12]. However, questions about the sustainability of bioenergy pathways have been raised, because the conversion of biomass into energy consists of input and output flows which may affect its environmental performance. Therefore, a holistic and comprehensive environmental tool like Life Cycle Assessment (LCA) can be used to assess and ensure the environmental sustainability of biogas supply chains [13,14].

Life cycle assessment (LCA) is a comprehensive assessment and a holistic approach that can provide relative and accurate information to be applied in environmental management [15–17]. LCA can be used to assess environmental burdens related to a product, process or service by identifying the energy, materials used and emissions released to the environment [18–20]. The LCA approach has long been practiced around the world, and it has significantly improved in recent years. According to Talve [21], LCA was first introduced in 1960 in the United States of America by Harold Smith. He presented his research in the 1963 World Energy Conference on the assessment of the cumulative energy needed to produce chemical products. However, this methodological approach is still new and under development in Malaysia. Hence, this study has taken a further step in the direction of evaluating the environmental impacts of biogas production based on the LCA perspective.

2. Sustainable Biogas Production

Recent years have seen a surge of interest in assessing the environmental impacts of the production of green goods and services. This is true particularly in the renewable energy system, increasingly concerned with sustainable environmental requirements. For example, Malaysia has recently become one of the most important poles of biofuel technology in the world [22]. This is due to its huge palm tree plantations, a source of biofuel production. Moreover, there is an abundant biomass source in Malaysia from agricultural crops and wastewater from industrial activity that can be utilized as feedstock for bioenergy production [23,24]. However, without appropriate wastewater treatment and management, a huge source of renewable energy will be wasted and, at the same time, become a menace to the environment. Hence, biogas production from available biomass waste and wastewater could be one of the suitable solutions to overcome the wastage. Biomass waste utilization has a direct impact on the recovery of energy. There are a number of energy recovery methods which can be used, such as biochemical (e.g., anaerobic digestion, composting and vermicomposting), thermal conversions (e.g., gasification, incineration, fast and slow pyrolysis) and chemical conversions (e.g., transesterification) [25,26].

In Malaysia, LCA studies were mostly conducted by the Malaysian Palm Oil Board (MPOB) and SIRIM Berhad. Various areas have been covered in the LCA study, such as waste management, petroleum, agro-industry and palm oil [27,28]. A study conducted by Aziz et. al [24] revealed the potential of biogas production from six types of substrates in Malaysia. In addition, recent studies on biogas production from the anaerobic digestion of a palm oil mill effluent have been conducted to highlight the feasibility of the LCA approach in biogas production, as well as the opportunity and challenges from the Malaysian perspective [27]. However, there are no strict regulations issued by the government concerning biogas plant installation and utilization, despite the various green policies that have been developed and introduced. Even though biogas generation is still at a nascent stage in Malaysia, it has a high potential in the way forward to achieve sustainable development. Therefore, it is important to assess the environmental performance of the system to ensure and enhance its sustainability [28,29]. A proper guideline, like a LCA-based conceptual framework, would also assist the government and related stakeholders in making decisions to improve the environmental performance of biogas production. Hence, an LCA-based conceptual framework was developed and proposed in this study (Figure 1). The framework shows the integration of policy drivers, proposed actions, existing green initiatives, green market influences and sustainability evaluation using the LCA approach affecting the sustainability of biogas production supply chains.

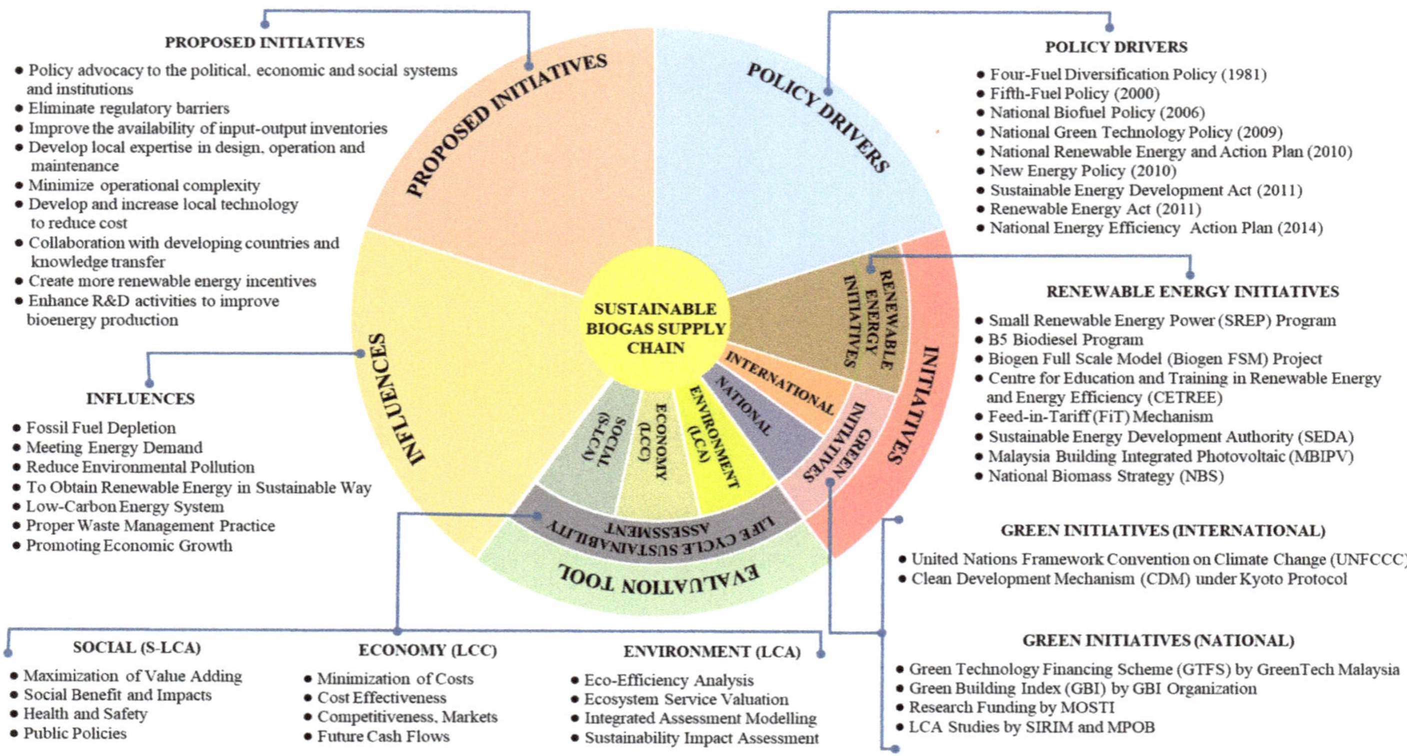

Figure 1. Conceptual framework of a life cycle assessment (LCA)-based environmental sustainability evaluation of biogas production supply chains.

The LCA-based environmental evaluation is an integrated approach which can assess the environmental performance of biogas production. LCA is a cradle-to-grave approach for assessing the impacts of a product throughout its life cycle, from raw materials extraction, through production, manufacture, transportation and use, to the management of the discarded product, either by recycling or final disposal [30]. The information from the conceptual framework of an LCA-based environmental evaluation of biogas production could assist the consumer in making choices towards green goods and services. Moreover, an LCA approach will act as a continuous measure and help both policy and decision makers to identify the opportunities for environmental improvements.

2.1. Policy Drivers

Many initiatives have been proposed and implemented by the government to promote sustainable development. As shown in Figure 1, a total of nine policy drivers have been identified concerning renewable energy evolution: Four Fuel Diversification Policy (1981), Fifth Fuel Policy (2000), National Biofuel Policy (2006), National Green Technology Policy (2009), National Renewable Energy and Action Plan (2010), New Energy Policy (2010), Sustainable Energy Development Act (2011), Renewable Energy Act (2011) and National Energy Efficiency Action Plan (2014). The energy-related policies were introduced to ensure future energy security and stability [31,32].

In 1981, the Four Fuel Diversification Policy was proposed to prevent over-reliance on oil as the main energy source, with a diversification of the energy mix to include gas, hydropower and coal. The National Biofuel Policy promoted the use of biofuels through incentives and making available 5% diesel and 5% palm olein biodiesel blends. The National Green Technology Policy targeted the development of renewable energy for energy security and considered renewable energy as an important factor for economic growth. The utilization of indigenous renewable resources was introduced through the National Renewable Energy and Action Plan to achieve electrical supply security and sustainable socioeconomic development. Under the Renewable Energy Act, a tariff system to promote renewable energy generation was established. The transition of attention towards renewable energy generation can be seen through the continuous development and realignment of energy policies.

Many countries' governments around the world have committed to decreasing greenhouse gas (GHG) emissions by promoting renewable energy. Both developed and developing countries have set a renewable energy target and promulgated legislation and regulations to encourage renewable energy development. Table 1 shows the regulations and measures related to renewable energy in several countries. The policies and measures were designed to reduce fossil fuel dependency, to promote the development and utilization of renewable energy, diversifying energy supplies, ensuring energy security, protecting the environment and considering economic and social sustainability [33,34]. According to Yaping Hua et al. [34], by the year 2013, at least 144 countries had set different renewable energy targets and policies at the national level. In addition, there was a total investment of 244 billion USD in renewable energy development globally.

Table 1. Renewable energy related policies and measures in several countries (adapted from [35]).

Country	Legislation and Regulation
	Developed Countries
Germany	2017 Amendment of the Renewable Energy Sources Act (EEG 2017) Subsidy for solar PV with storage installations Ground-Mounted PV Auction Ordinance (2015) 2014 Amendment of the Renewable Energy Sources Act (EEG 2014) CHP Agreements with Industry KfW Program Offshore Wind Energy Law on Energy and Climate Fund "Energy of the Future" monitoring process Sixth Energy Research Program

Table 1. *Cont.*

Country	Legislation and Regulation
	Developed Countries
Germany	Biofuels Quota Act (2010) Energy Concept National Energy Action Plan (NREAP) (2010) KfW Renewable Energies Program KfW Program Energy-Efficient Rehabilitation Renewable Energies Heat Act (2009) Climate Legislation Package Enacted under the Integrated Climate Change and Energy Program (2008) Integrated Climate Change and Energy Program Funding for Solar Power Development Center Energy Industry Act (2005) (amended 2012) Law to Amend the Mineral Oil Tax Law and Renewable Energy Law (2002) Combined Heat and Power Law (2002) Eco-Tax Reform (1999) Market Incentive Program Preferential Loan Program offered by the Reconstruction Loan Corporation (KfW) Federal Building Codes for Renewable Energy Production Green Power Ordinance on the Fee Schedule for Architects and Engineers (1995) Federal States Support for Renewable Energy (1985)
Australia	Renewable Energy (Electricity) Act 2000 Renewable Energy (Electricity) (Small-Scale Technology Shortfall Charge) Act 2010 Renewable Energy (Electricity) (Large-Scale Generation Shortfall Charge) Act 2000 Renewable Energy (Electricity) Regulation 2001, Renewable Energy Certificates (RECs)
Japan	Law Concerning the Promotion of the Development and Introduction of Alternative Energy Long-Term Energy Supply and Demand Outlook (2015) Strategic Energy Plan (2014) Feed-in Tariff for Renewable Electricity and Solar PV Auction (2012) Global Methane Initiative (2010) Cool Earth-Energy Innovative Technology Plan (2008) Seaway Signals Converted to Use Renewable Energy (2000) Promotion of New and Renewable Energy (1997) Special Measures Law for Promoting the Use of New Energy (1997) Projects for Development and Deployment of New and Renewable Energy by NEDO and by NEPC (1980) Law on Establishment of NEDO (1980)
	Developing Countries
China	Renewable Energy Law Regulation on Administration of Power Generation from Renewable Energy Measures on Supervision and Administration of Grid Enterprises in the Purchase of Renewable Energy Power Trial Management Measures for Renewable Power Pricing and Cost Share Trial Management Measure for Allocation of Renewable Energy Tariff Surplus Revenue Notice of Strengthening the Construction and Management of Biofuel Ethanol and Promoting Sound Industrial Development Trial Management Measures for the Special Development Fund Implementation Guidelines on Promoting Wind Power Industry Guides to Renewable Energy Development
Thailand	Thailand Alternative Energy Development Plan (AEDP 2015-2036) Feed-in Tariff for Very Small Power Producers (VSPP) (excluding solar PV) Feed-in tariff for distributed solar systems Biodiesel blending mandate Renewable Energy Development Plan (REDP) 2008-2022 Small and Very Small Power Purchase Agreements Energy Conservation Program (ENCON)

Table 1. *Cont.*

Country	Legislation and Regulation
	Developed Countries
Vietnam	National Power Development Plan 7 (PDPD7 – revised) (2016) Vietnam Renewable Energy Development Strategy 2016-2030 with outlook until 2050 (REDS) Decision on support mechanisms for the development of biomass power project in Vietnam (biomass feed-in tariff) Decision on support mechanisms for the development of waste-to-energy power projects in Vietnam (feed-in tariff) Accelerated depreciation tax relief for renewable energy projects Electricity Law (2005) Decree No. 45/2001/ND-CP on electric power operation and use (2001)

2.2. The Green Initiatives in Malaysia

The government has provided a great effort and commitment through many renewable energy and green initiatives at both the national and international levels to promote green and sustainable development (Figure 1). Renewable energy initiatives like the Small Renewable Energy Power (SREP) program, Biogen Full Scale Model (Biogen FSM) project, Feed-in-Tariff (FiT) mechanism, Malaysia Building Integrated Photovoltaic (MBIPV), B5 biodiesel program, Centre for Education and Training in Renewable Energy and Energy Efficiency (CETREE), Sustainable Energy Development Authority (SEDA) and National Biomass Strategy (NBS) have been introduced [36,37]. In addition, the Green Technology Financing Scheme (GTFS) by GreenTech Malaysia, Green Building Index (GBI) by GBI organization, research funding by MOSTI and LCA studies by SIRIM and MPOB are the examples of green initiatives conducted at the national level.

Malaysia also took part at the international level (i.e., United Nations Framework Convention on Climate Change (UNFCCC) and Clean Development Mechanism (CDM) under the Kyoto Protocol), which shows the commitment of the government to shifting towards cleaner and sustainable development. More innovative initiatives and strategies would help to identify multiple pathways towards sustainable energy. Due to this commitment, the environmental evaluation of a biogas production system using LCA tools can be integrated into Malaysia's policies and action plans to take further steps towards sustainable development in the future.

2.3. Life Cycle Sustainability Assessment

Sustainable development is defined as development that meets the needs of the present without compromising the ability of future generations to meet their own needs [38]. In order to develop a life cycle sustainability assessment (LCSA), integrated environment, economy and social perspectives need to be considered by looking at three different dimensions of the same system [39–41] (Figure 1). The LCA produces numerical data and indicators to evaluate the used resources and environmental impacts; life cycle costing (LCC) produces cost indicators to evaluate the cost-effectiveness; and the social life cycle (S-LCA) introduces social indicators to evaluate the corporate policy and human rights of a product or system. Therefore, if the study could extend into economic and social dimensions in the future, a thorough understanding on the sustainability of the system could be achieved. According to Coyle and Rebow [42] and Jürgensen et al. [43], sustainability in energy systems is usually associated with energy efficiency and energy with lower emissions. An evaluation tool like the LCA can be used to assess the environmental performance and determine the hotspots along the supply chains, thus providing a comprehensive assessment of the sustainability of the renewable energy system. The evaluation tool is one of the important factors that should be integrated into sustainable energy, in addition to public awareness, sustainability education and training and the promotion of renewable energy resources.

The environmental life cycle assessment of biogas production supply chains helps to define the up- and down-streams of the whole system and identify possible problems at each stage of the evaluation process [44,45]. The contribution of the process to the selected impact categories can be determined. Several actions and improvement strategies can be proposed for enhancing the environmental sustainability of biogas production. Other tools that can be used to support the decision making and to evaluate the impacts of renewable energy systems are the ecological footprint, energy analysis, net energy balance, carbon footprint, GHG life cycle analysis, material flow analysis, sustainability indicators, fuel cycle analysis and life cycle risk assessment [46]. However, it has been found that the LCA is preferable as a research-based approach and able to holistically assess the environmental sustainability of renewable energy systems [46]. As reported by Sala et al. [47] and Takeda et al. [48], the LCA approach could identify deficiencies for further improvements and help to avoid problem and burden shifting from one part of the system to the others.

Bioenergy production from organic waste does not automatically imply that its production, conversion and utilization are sustainable [49]. Based on the findings of LCA studies for renewable energy by Milazzo et al. [50] and Turconi et al. [51], the sustainability of bioenergy is more than just GHG savings. It is also associated with specific resource use and the potential environmental consequences. The sustainability of these components is important because any deficiencies could lead to market distortion and consequently impede the efforts to increase renewable energy shares in Malaysia's energy capacity.

2.4. *Influences and Proposed Actions*

Figure 1 shows some issues that influence the growing interest towards sustainable renewable energy production. Fossil fuels depletion, increasing energy demand and environmental pollution problems have been a major influence in the transition of attention towards renewable and sustainable energy [52,53]. The current plan also aims to obtain renewable energy in sustainable ways, to have a low-carbon energy system and to have a proper waste management practice, as well as promoting economic growth. However, it is crucial to have an appropriate action plan to overcome the obstacles and barriers which limit the feasibility of sustainable biogas production supply chains in Malaysia.

According to Figure 1, a few strategies and initiatives that could be proposed to help improve the sustainability of the system are the research and development (R&D) activities, which should be enhanced to improve bioenergy production, more renewable energy incentives, collaborations with developing countries and knowledge transfer, developing and increasing local technology to reduce cost, minimizing operational complexity, developing local expertise in design, operation and maintenance, improving the availability of input-output inventories, eliminating regulatory barriers and also conducting policy advocacy to the political, economic and social systems and institutions.

3. Promoting Biogas Production from a Palm Oil Mill Effluent as a Renewable Energy Source in Malaysia

The palm oil industry in Malaysia has boosted the country's economy, as it was one of the largest palm oil exporters and producers in the world. The palm oil industry has supported a variety of food products like margarine, cooking oils and animal feeds, and non-food products like soaps, detergents, cosmetics, pharmaceuticals and biofuels [54–58]. The by-products generated during the palm oil milling process can be recycled and recovered into reusable materials or products (e.g., shells and fibers used as fuel for boilers, empty fruit bunches used as soil conditioner and palm oil mill effluent (POME) utilized as a source of energy generation) [59,60]. Figure 2 shows the overview of the flow process and products of the palm oil industry. According to Basri et al. [61] and Oswal et al. [62], POME has been applied as feedstock in most palm oil mills to generate biogas. Due to its high organic content, POME can be a good source to generate methane gas for energy production. The exploitation of POME for renewable energy production would enhance the sustainability of palm oil industries. As a consequence, bioenergy generation from POME can be an added value to the palm oil industry in Malaysia.

Issues on energy security and environmental concerns lead to the utilization of renewable energy sources. Upgrading biogas to biomethane and injecting it into the natural gas grid could be an efficient way of integrating the biogas into the energy sector [60]. Accordingly, biomethane can substitute the use of natural gas and can also be used as transportation fuel. However, the unattractive connection price to the grid, the irregular supply of biomass, the low efficiency of combustion technology and poor supporting systems like interconnection infrastructure have caused a shortfall in national renewable energy capacity in Malaysia. In 2017, the total electricity demand in Malaysia was approximately 12 607 ktoe [27]. The estimated energy potential generated from 50k tonnes of POME produced in a year is about 3.2 million MWh of electricity, contributing to 2.19% of the total electricity demand. Though the utilization of biogas as renewable energy is still unregulated in Malaysia, it is crucial because it could improve long-term energy security and environment protection.

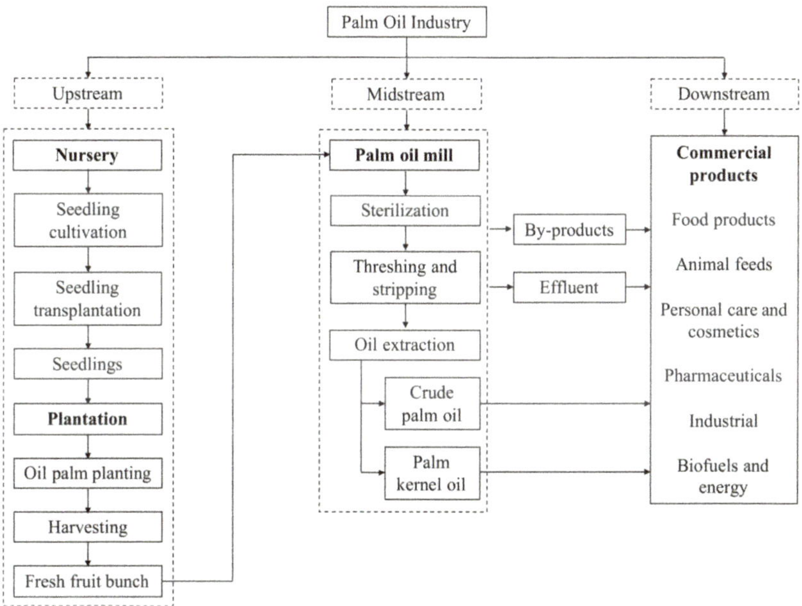

Figure 2. Overview of the flow process and products of the palm oil industry in Malaysia.

Current developments in the local and global economies are closely connected to sustainable energy resources. The feed-in-tariff (FiT) mechanism administered by the Sustainable Energy Development Authority (SEDA) was designed to correspond to the National Renewable Energy and Action Plan (2010), that suggested the requirement of legislative solutions to increase the renewable energy share in Malaysia's energy mix [63,64]. The type of resources included in the FiT mechanism are biogas, biomass, hydropower and solar photovoltaic power. At 40 MW, the oil palm biomass contributed the highest grid-connected capacity among renewable technologies in the Ninth Malaysia Plan (2006–2010) [65–67]. In addition, oil palm biomass emerged as a major contributor towards achieving the capacity target set under the FiT of 800 MW of grid-connected capacity by the year 2020 [63,68].

However, the survey carried out by Umar et al. (2014) [67] showed that some of the palm oil millers were less interested in embarking on a renewable energy venture. This is due to the fact that most of the mill operators' financial capability is too low to participate in a high cost project, and renewable energy was not a profitable business. Therefore, an innovative incentives scheme should be proposed to reduce the financial pressure on energy providers. In addition, the cost could be reduced by developing and increasing local technology, as well as local expertise in design, operation and maintenance. The operational complexity should also be minimized, so that the system can be conducted and managed easily and the palm oil millers would not overlook the efficiency and effectiveness of the system. Other strategies that could be executed to enhance and promote the sustainability of biogas production in the country are the improvement of research and development (R&D) activities and the establishment of collaborative research with other developed and developing countries, along with a concurrent transfer of knowledge.

Conducting an LCA for a product or system requires a transparent inventory, so that an excellent overview of areas in which inputs could be substituted by less polluting materials can be acquired for the good of the environment. Hence, it is important for industries to record and make available their input and output inventories for research purposes. The conceptual framework of the LCA-based environmental evaluation developed in the present study can be used as a guideline for related stakeholders and decision makers to improve the quality and enhance the environmental sustainability of biogas production from POME in Malaysia, as well as promoting biogas as a cleaner, reliable and secure energy. The stakeholders include the government, oil palm producers, financial institutions and oil palm associations. Government intervention is crucial to drive the industry forward to achieve cleaner technology and towards sustainable development. Accordingly, the market community also shares an equal responsibility and influence, in order to increase the use of renewable energy sources and meet the sustainable development goals. A policy advocacy of sustainable renewable energy resources should be conducted at the political, economic, social and institutional levels, to raise awareness to the fact that environmental-friendly products and systems are starting to get local and global attention. According to Aziz et al. [69], various studies on the implementation of the LCA approach to evaluate the environmental performance of biogas production show that the LCA could help improve the environmental profile of the biogas system. There were many initiatives by the Malaysian government towards sustainable development, which can be a potential driving force to apply the LCA to the environmental evaluation of a clean technology.

4. Life Cycle Assessment of Zero Discharge Treatment

POME is a remarkably contaminating effluent because of its high content of organic matter (expressed as chemical oxygen demand (COD) and biochemical oxygen demand (BOD)), which can have harmful effects on the environment, particularly on the water resources to which POME is discharged. In recent years, POME has been recognized as a prospective source of renewable energy. Accordingly, the production of bioenergy will be more sustainable and cleaner when operating simultaneously with wastewater treatment. There are various existing processes for POME treatment and the conversion of POME into bioenergy, such as aerobic and anaerobic digestion, physico-chemical treatment and membrane separation [70–73]. Nevertheless, in the integrated biological treatment of the POME system, which includes membrane treatment, higher quality effluents can be produced. This system is also called zero discharge treatment system. In the zero waste discharge treatment system, all wastewater is purified and recycled. Therefore, the plant discharges zero effluents into the water, which in effect reduces environmental pollution. The anaerobic digestion of POME can produce biogas and a final discharge which meets the proposed regulatory standard of the Department of Environment (DOE), Malaysia. However, in the zero discharge treatment system, biogas can be produced with a maximum water recycle and reuse.

Zero waste discharge treatment has been implemented in several countries around the world (i.e., Italy, United States, Canada, Malaysia, China and India). For instance, this technology has been used for reusing the wastewater produced by car manufacturing and secondary sewage [74], heavy oil recovery [75], wastewater reclamation and reuse in marine ports [76], the treatment of chromium-containing leather waste [77] and treating POME [78]. However, the environmental sustainability of this treatment process in terms of the LCA approach is still under development. One study has been carried out in India by Rajakumari and Kanmani [79] on zero liquid discharge treatment technologies for textile industries, using pretreatment, reverse osmosis (RO) and evaporator treatment units. The results of the study show that the energy consumption in the treatment plant had contributed to the global warming potential (GWP) by

5.56 kg CO_2-eq per functional unit of 1 m^3 of textile wastewater. It was reported that CO_2 was a leading pollutant, which can be reduced by using biomass gasification.

Wastewater recovery and recycling has become a growing trend because it not only minimizes the environmental impact of discharged water, but can also be an additional water resource which promotes water sustainability [80]. On that account, palm oil mills in Malaysia, particularly, should explore and adopt the zero discharge technology to achieve zero discharge concepts. According to Madaki and Seng [81], POME is a very difficult and expensive waste to manage. Hence, there is a need to shift from the conventional system to the current advanced POME treatment technology. Although the zero discharge treatment system is a promising sustainable technology, its application requires a high amount of energy consumption, which will result in some environmental impacts [82]. Therefore, an LCA of the zero waste discharge treatment process is important, as it will provide additional insights into the environmental sustainability of the technology.

4.1. Material and Methods

4.1.1. Goal and Scope, Unit Process, Functional Unit and System Boundaries

In this study, the zero discharge treatment of POME has been used as a case study. The LCA includes four phases: goal definition and scope, life cycle inventory (LCI) analysis, life cycle impact assessment (LCIA) and interpretation. A gate-to-gate approach was used, focused on the process of zero discharge treatment. The goal of using the LCA method was to determine the environmental impacts of the zero discharge treatment system and to derive measures to reduce them. The energy and material requirements, as well as the emissions to the environment, were taken into account. The functional unit for impact assessment was a metric tonne of POME. Figure 3 shows the system boundary for the LCA of the zero discharge treatment of POME. Three main processing units (i.e., pretreatment, biological treatment (anaerobic and aerobic treatment) and post-treatment (ultrafiltration and reverse osmosis)) were included in the environmental assessment.

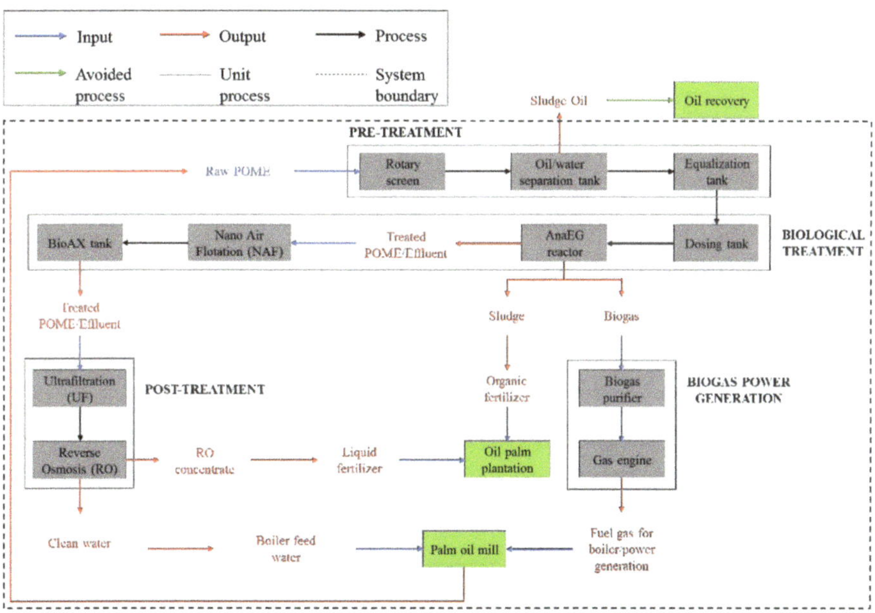

Figure 3. Life cycle flow chart of a zero discharge treatment system of POME.

4.1.2. Database and Analysis Methods

SimaPro 8.5.2 software was used to gather and analyze the inventory data. SimaPro 8.5.2 is an LCA software that can be used to monitor the performance of the sustainability of a product or service. This software can analyze a complex life cycle systematically and can evaluate the environmental impact of a product or service at each stage of the life cycle. Ecoinvent 3.4 [83], Agri-footprint 4.0 [84] and USLCI [85] were chosen as background data sources. Ecoinvent 3.4 database was used for this study because it contains LCI data from various sectors, such as energy production, transportation, chemicals production and also fruits and vegetables. Agri-footprint 4.0 database was chosen due to its comprehensive LCI database focusing on the agriculture and food sectors, which covers data on agricultural products (i.e., food, feed and biomass). Besides, Agri-footprint 4.0 covers materials and inputs from agriculture, while USLCI database contains data modules that quantify the material and energy flows into and out of the environment. Using Ecoinvent 3.4, Agri-footprint 4.0 and USLCI data as proxies, the background databases were already incorporated in SimaPro, while foreground databases were newly included in the software for analysis, obtained from survey and existing regional datasets.

Life cycle impact category indicators were calculated using ReCiPe 2016, developed by Huijbregts et al. [86]. ReCiPe 2016 evaluates 18 different impact categories at the midpoint level, namely global warming (GWP), stratospheric ozone depletion (ODP), ionizing radiation (IRP), ozone formation (human health) (HOFP), fine particulate matter formation (PMFP), ozone formation (terrestrial ecosystems) (EOFP), terrestrial acidification (TAP), freshwater eutrophication (FEP), marine eutrophication (MEP), terrestrial ecotoxicity (TETP), freshwater ecotoxicity (FETP), marine ecotoxicity (METP), human carcinogenic toxicity (HTPc), human non-carcinogenic toxicity (HTPnc), land use change (LUC), mineral resource scarcity (SOP), fossil resource scarcity (FFP) and water consumption (WCP). The impact categories were then divided into three damage assessment categories at the endpoint level, namely damage to human health (HH), damage to ecosystem quality (ED) and damage to resource availability (RA).

4.1.3. Life Cycle Inventory

The life cycle inventory includes inventory data from the inputs and outputs in the process of zero discharge treatment. Inventory data consist of the amount of energy and materials consumed and the quantities of emissions released to the environment. All data were obtained from a set of questionnaires, on-site data, literature reviews and databases in SimaPro 8.5.2. The raw data information was obtained directly from the owner of the mill chosen in the present study. For compatibility, the questionnaires were developed based on the guideline provided by ISO 14044 [87], consisting of the generic information (i.e., annual reports and company brochures), process description and the input and output flows of the analyzed product. An interview with the mill owners and officers was conducted to obtain and validate information regarding the production process, technology used, mill operation and operational issues.

The case study was based on POME-based biogas production from a palm oil mill with biogas and zero discharge treatment facilities in Malaysia, and the data were collected for a period of three years. Data on the inputs and outputs of the production process, including the amount of energy, materials consumed and transportation, were available from 2012 to 2016. However, the quantity and quality of the data were not consistent from year to year. Therefore, this study was conducted based on the average of three years (2013 until 2015) data. The data chosen were the most complete and with best quantity and quality. The validation of data was performed by on-site visits, comparison with other data sources and recalculation. The results from the LCI were subsequently used to assess the environmental performance of the zero discharge treatment process.

The zero discharge treatment system of POME consists of three main processing units, as shown in Figure 3: pretreatment, biological treatment and post-treatment. The core technologies in the POME zero discharge system in this study were AnaEG and BioAX. AnaEG is an advanced anaerobic expanded granular sludge bed reactor, while BioAX is an advanced biocontact aerobic process with internal circulation. In the AnaEG bioreactor, the wastewater flows upwards in a plug flow pattern consisting of a two-phase anaerobic process in one reactor. The BioAX tank contains biofilm that provides a suitable environment for the growth of microbes to further enhance the degradation of organic matter. In BioAX, returned sludge was not required. Table 2 shows the inventory data for the treatment of a metric tonne of POME. The treatment plant was installed in the palm oil mill, and hence no vehicle was needed for the transportation of feedstock. In this study, only a small amount of water is required for the treatment process, approximately 0.2 L per tonne of POME. Rainwater harvesting has been practiced in the treatment plant to collect and store rainwater for on-site reuse. Though Malaysia has sufficient water resources, some regions in the country are facing water scarcity, and the water demand is also increasing nowadays. Hence, rainwater harvesting as an alternative water resource was proposed by the government as part of the solutions to mitigate water scarcity problems [88]. 50% of the total electricity consumption was taken from the national grid and another 50% from the biogas engine. The utilization of biogas for electricity generation could reduce the cost as well as the dependency on non-renewable energy sources.

Table 2. LCI for the zero discharge treatment system of POME (per tonne of POME).

Input/Output	Unit	Amount	Data Source	Link Process/Substance in SimaPro 8.5.0
Input from technosphere				
Materials and fuels				
Palm oil mill effluent (POME)	t	1	On-site data	Palm oil mill effluent
Energy				
Electricity from grid	kWh	0.385	On-site data	Electricity mix, AC, consumption mix, at consumer, < 1 kV/MY Mass
Electricity from biogas	kWh	0.321	On-site data	Biogas, zero discharge treatment plant/MY
Output to technosphere				
Products and co-products				
Clean water	t	0.75	On-site data	Clean water
Biogas	m^3	28	On-site data	Biogas, zero discharge treatment plant/MY
Sludge oil	kg	4.70	On-site data	Sludge oil
Organic solid sludge	kg	28.98	On-site data	Organic solid sludge
RO concentrate (K_2O & MgO)	kg	3.41	On-site data	RO concentrate (K_2O & MgO)
Input from environment				
Resource				
Water	m^3	0.0002	On-site data	Water, rain

During the pretreatment of POME, suspended matter such as oil and solids was removed to ensure the effective anaerobic treatment of POME. The sludge oil produced was recovered and sold out as by-product. The biological treatment process produced an average of 28 m^3 biogas per tonne of POME and generated a final discharge with a BOD of less than 20 mg/L. The biogas composition was 65-70% CH_4, 25-36% CO_2 and 800-1500 ppm H_2S. The biogas produced was burned in a gas engine to generate power for internal use as electricity and was also utilized as fuel for the boiler to generate steam in the palm oil mill. The treated sludge (digestate) was recovered from the anaerobic digester and sent to the plantation as an organic fertilizer. According to Loh et al. [89], the application of the treated sludge as organic fertilizer could enhance soil fertility due to its high content of nutrients. In this treatment system, no water back flow process is needed in the anaerobic reactor and no returned sludge is required in the aerobic treatment process. The effluent generated from the biological treatment contains a low level of organic and suspended solids. Finally, the post-treatment process' final discharge was clean water that can be reused as boiler feed water in the mill. An RO concentrate amounting to 40% of rejected water was recovered and collected as liquid fertilizer with s high content of potassium and magnesium.

The ultimate goals of this treatment system are to produce biogas as a source of renewable energy, zero emissions of GHG, a final discharge with less than 20 mg/L of BOD and clean water which can be used as boiler feed water. Table 3 shows the characteristics of POME after each treatment in the zero discharge treatment system according to Tabassum et al. [90]. In the presented results, the average removal of COD and BOD after the biological treatment were 98.5% and 99.9%, respectively. After the membrane treatment UF/RO process, the value of COD and BOD were almost undetectable. The quality of water from each treatment stage shows an improvement in terms of color, odor and turbidity and obtained > 90% removal of chemical oxygen demand (COD), biochemical oxygen demand (BOD) and suspended solid (SS). At the final stage, the RO permeate was odor-free, clear and with a pH of 9.48. The zero discharge treatment system was able to reclaim water as boiler feed water, since the water quality complied with the boiler feed water standard set by the American Boiler Manufactures Association (ABMA), which is pH 7.5 until 10 [78]. It is the ideal pH for boiler feed water to prevent corrosion. After the entire treatment processes, higher percentages of water could be recovered and used as boiler feed water, and the recovered biogas could be utilized to produce heat and electricity for self-consumption. Therefore, the products and effluents produced can be considered environmentally sustainable.

Table 3. The characteristics of POME after each treatment (adapted from [80]).

Parameter	Unit	Raw POME	EQ	AnaEG Effluent	Nano Clarifier	BioAX Effluent	UF/RO Permeate
pH		4.30	4.00	7.00	7.50	8.00	9.48
COD	mg/L	75000	65000	4500	2000	1100	ND
BOD_5	mg/L	27000	NM	NM	820	<20	ND
TS	mg/L	100000	48600	22600	8200	5650	NM
SS	mg/L	50000	22778	13840	350	191	ND
TVS	mg/L	80000	40200	14300	3000	1600	NM
Dissolved solids	mg/L	50000	25882	8760	7850	5459	NM
VFA	mg/L	2184	NM	413	NM	NM	NM
Total Alkalinity	mg/L	536	NM	4100	NM	NM	NM
Turbidity	NTU	NM	NM	NM	700	110.0	0.4

ND: not detected; NM: not measured.

4.2. Results and Discussion

Life Cycle Impact Assessment

The potential impacts were determined using the results from the inventory analysis (LCI). There are four steps in the life cycle impact assessment (LCIA): classification, characterization, normalization and weighting. According to ISO 14044 [87], classification and characterization are mandatory steps for the LCIA, while normalization and weighting are optional steps depending on the goal and scope of the study. In the present study, the classification and characterization steps were included.

The assessment has been carried out on a zero discharge treatment system based on the functional unit of 1 tonne of POME. The impact results at the midpoint level are presented as characterization values in Table 4. The characterization factors indicate the environmental impact per unit of stressor. As presented in Table 4, the most significant contribution to all impact categories originated from the electricity consumption from the national grid. Hence, electricity production from the national grid is an important contributor to all impact categories. This is because the Malaysian grid energy profile is largely composed of non-renewable energy sources, with 45% from natural gas, 41% from coal, 6% from hydroelectric power and 8% from oil [91].

Table 4. The characterization values at the midpoint level per 1 t of POME.

Impact Category	Unit	Total
GWP	kg CO_2-eq	4.42×10^{2}
ODP	kg CFC11-eq	7.75×10^{-9}
IRP	kBq Co-60-eq	4.80×10^{-7}
HOFP	kg NO_x-eq	2.15×10^{-4}
PMFP	kg $PM_{2.5}$-eq	1.49×10^{-3}
EOFP	kg NO_x-eq	2.17×10^{-4}
TAP	kg SO_2-eq	1.00×10^{-2}
FEP	kg P-eq	6.65×10^{-12}
MEP	kg N-eq	2.42×10^{-7}
TETP	kg 1,4-DCB	3.08×10^{-2}
FETP	kg 1,4-DCB	1.82×10^{-4}
METP	kg 1,4-DCB	2.49×10^{-4}
HTPc	kg 1,4-DCB	7.93×10^{-5}
HTPnc	kg 1,4-DCB	8.50×10^{-3}
SOP	kg Cu-eq	1.32×10^{-7}
FFP	kg oil-eq	4.06×10^{-2}
WCP	m^3	2.02×10^{-4}

In 2015, the total electricity generation in Malaysia was approximately 144,565 GWh, and 89.3% of the electricity mix was generated by using fossil fuels [92]. The energy demand in Malaysia had increased from 40,845 ktoe in 2009 to 51,807 ktoe in 2015 [93]. It was reported that the use of energy for electricity generation is the largest source of emissions of GHG when compared to other human activities. From 2000 to 2015, the emissions from electricity generation had increased by 45% [91]. CO_2 emissions resulting from the oxidation of carbon in fuels during combustion account for the largest share of global anthropogenic GHG emissions. The cumulative effect of other GHG emissions like CH_4 and N_2O towards global climate were estimated to be at least one order of magnitude lower than that of CO_2 [94,95]. However, fossil fuels are still being utilized extensively by developing countries in order to meet the energy demand and to support development and economic growth.

In this study, the treatment plant consumed approximately 0.38 kWh of electricity from the national grid, which affected 17 impact categories. The impacts on GWP, ODP, HOFP, PMFP, EOFP, TAP, MEP, TETP, FETP, METP, HTPc, HTPnc and FFP were caused by the production of natural gas, anthracite coal, lignite coal, bituminous coal, diesel and residual fuel oil. The electricity production from hydroelectric power contributed to IRP, FEP, SOP and WCP. As reported before, the electricity generation in Malaysia includes production processes from several fossil fuel resources. The burning of fossil fuels has huge consequences for the environment. Figure 4 shows the proportion of electricity production resources which contributed to 13 impact categories. The data for the electricity production process were adopted from the USLCI and ELCD databases. In the presented results, natural gas was the largest proportion contributing to the GWP, ODP, MEP, FETP, METP, HTPnc and FFP impact categories, at 37% (4.39×10^{-2} kg CO_2-eq), 91% (7.03×10^{-2} kg CFC11-eq), 76% (1.83×10^{-7} kg N-eq), 66% (1.20×10^{-4} kg 1,4-DCB-eq), 60% (1.51×10^{-4} kg 1,4-DCB-eq), 43% (3.64 × 10–3 kg 1,4-DCB-eq) and 40% (1.62×10^{-2} kg oil-eq), respectively. Anthracite coal was the largest contributor towards PMFP, TAP, TETP and HTPc, at 51% (2.30×10^{-4} kg $PM_{2.5}$-eq), 51% (7.87×10^{-4} kg SO_2-eq), 68% (2.09×10^{-2} kg1,4-DCB-eq) and 47% (3.71×10^{-5} kg 1,4-DCB-eq), respectively. For HOFP and EOFP, the largest contributor was lignite coal, with an impact of 31% (6.61×10^{-5} kg NO_x-eq) for both categories. For electricity production from hydropower, which is a non-fossil energy, the assumption is made that there is no environmental impact.

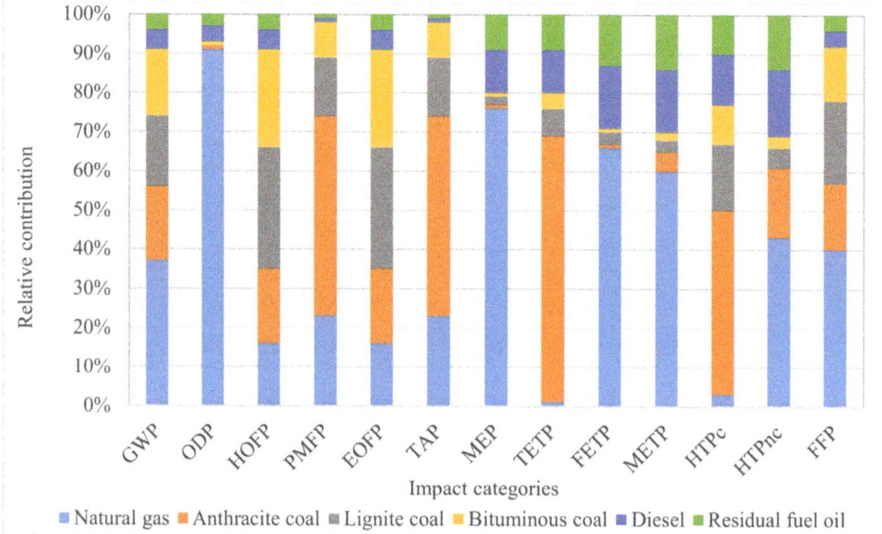

Figure 4. Relative contribution from electricity sources to each impact category.

Based on the study conducted by Lijó et al. [96] on the LCA of electricity production from anaerobic co-digestion of pig slurry and energy crops in Italy, the electricity consumption from the grid contributed to the abiotic depletion potential (ADP), ozone layer depletion potential (ODP) and nuclear energy demand (CED). This is due to the fact that the Italian grid was also composed of non-renewable energy. According to Wang et al. [97], in the study on the LCA of biogas production from straw in China, the electricity consumption from the grid had a harmful effect on human health. The electricity was mostly produced from coal-fired power plants which release gaseous pollutants. Those findings were consistent with those of Roux et al. [98], which show that the impact of electricity consumption on GWP and ADP were due to the higher share of coal and gas power plants in the French electricity mix.

Table 5 shows the results of the damage assessment at the endpoint level. The zero discharge treatment system impacted HH, ED and RA at 4.13×10^{-4} DALY, 1.24×10^{-6} species·year and 0.009 USD, respectively, due to electricity consumption. As reported by Loh et al. [89], Tabassum et al. [90] and Wang et al. [99], there were no GHG emissions into the atmosphere from the zero discharge treatment system, as the system produces zero discharge. Nevertheless, this study shows, on the other hand, the pollution emissions in the system were caused by the electricity consumption. Although the treatment system could treat all the incoming effluent and leave nothing behind, there are other aspects that should be taken into consideration (i.e., energy sources, water sources, etc.) to ensure the environmental sustainability of the system holistically.

Table 5. The results of damage impact at the endpoint level.

Area of protection	Unit	Total
Human Health		
GWP, Human health	DALY	4.12×10^{-4}
ODP	DALY	4.11×10^{-12}
HOFP	DALY	1.96×10^{-10}
PMFP	DALY	9.38×10^{-7}
HTPc	DALY	2.63×10^{-10}

Table 5. *Cont.*

Area of protection	Unit	Total
Human Health		
HTPnc	DALY	1.94×10^{-9}
Total	**DALY**	$\mathbf{4.13 \times 10^{-4}}$
Ecosystem Damage		
GWP, Terrestrial ecosystems	species.yr	1.24×10^{-6}
GWP, Freshwater ecosystems	species.yr	3.39×10^{-11}
EOFP	species.yr	2.80×10^{-11}
TAP	species.yr	2.13×10^{-9}
MEP	species.yr	4.12×10^{-16}
TETP	species.yr	3.51×10^{-13}
FETP	species.yr	1.26×10^{-13}
METP	species.yr	2.62×10^{-14}
Total	**species.yr**	$\mathbf{1.24 \times 10^{-6}}$
Resource Availability		
FFP	USD	9.00×10^{-3}
Total	**USD**	$\mathbf{9.00 \times 10^{-3}}$

Figure 5 shows the relative contribution from electricity production resources to each damage impact category. Anthracite coal had the largest impact to both HH and ED categories, at 42.1% (1.66×10^{-7} DALY) and 34.4% (2.35×10^{-10} species.yr), respectively. According to Oh et al. [93], about 70% of coal demand is for energy production. The coal reserves in Malaysia were estimated to be approximately 1.9 billion metric tonnes [92]. The country consumed more than 20 million metric tonnes annually, and therefore 90% of coal supplies were imported mainly from Indonesia and Australia [100,101]. However, the combustion of coal poses major challenges like GHG emissions and air pollutants such as sulphur dioxide (SO_2) and CO_2. In contrast with gas, coal emitted a higher amount of CO_2 emissions due to its heavy carbon content per unit of energy released [102]. The emission of GHG will lead to an increase in the global mean temperature, which will consequently result in damages to human health and ecosystems. Therefore, at the endpoint level, damages to human health, terrestrial ecosystems and freshwater ecosystems can be estimated.

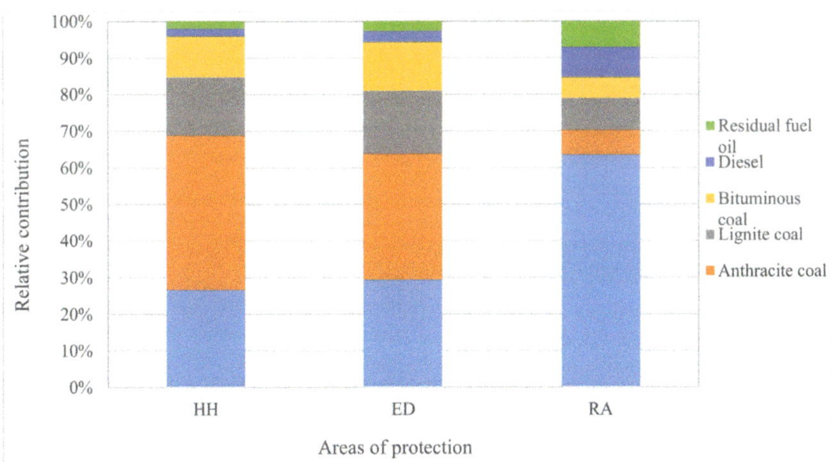

Figure 5. Relative contribution from electricity production resources to each damage impact.

On the other hand, the largest process contribution to the resource availability impact category was from natural gas, corresponding to 63.5% (5.78×10^{-3} USD). Natural gas, also known as liquefied natural gas (LNG), was the main contributor in Malaysia's energy mix [103]. At the endpoint level, the scarcity of fossil resources is expressed in economic terms [104]. The increase in fossil fuel extraction will cause an increase in costs. Hence, it will lead to a surplus cost potential (SCP), and the damage to natural resource scarcity can be estimated. In 2016, Malaysia became the world's third largest exporter of natural gas, after Qatar and Australia, with an estimated total natural gas reserve of 100.7 tscf [105]. Natural gas production has increased from 54.8 Mtoe in 2010 to 61.4 Mtoe in 2015 to fulfill the growing domestic demand and export contracts [93,106]. The rising of development and exploration activities has subsequently put pressure on the natural gas supply, and more investment is needed for reservoir development.

Electricity was required to run the plant mainly in loading operations, in the digester and in the chiller. More treatment units were used in the zero discharge treatment plant (i.e., AnaEG reactor, BioAX reactor, Nano Air Flotation system and membrane filtration), which may have caused higher electricity consumption. With regard to the impact of electricity consumption, the environmental sustainability of the system can be improved by increasing or enhancing the energy produced from biogas. In the zero discharge treatment plant, the engine generator, also known as gen-set, was used to generate electricity. The generated electricity was utilized for self-consumption and not exported to the grid. Anyhow, most of the biogas plants in Malaysia were running with combustion engines, combined heat and power (CHP) systems or a combination of both technologies [107]. Decentralized power generation with CHP units is the common biogas utilization pathway. Normally, in the typical electricity generation technology (i.e., conventional electricity generation, on-site boiler), the energy produced is wasted in the form of heat discharge to the atmosphere [98,99]. Conventional energy systems include power plants using fossil fuels. By using the CHP generation system, the transmission losses and carbon emissions could be reduced [108–112].

As reported in the study by [67], 86% of the market community were utilizing their biomass resources for on-site consumption, while only a few exported their excess electricity to the grid. This is because the energy produced for on-site electricity generation was sufficient, but it was not sufficient for exporting to the grid. Also, there is a lack of grid transmission lines connecting mills to the existing network systems due to distance constraints [60]. During the Ninth Malaysia Plan (2006–2010), the capacity share of renewable energy in the country's energy mix was 350 MW, or 1.8%. However, due to the slow renewable energy projects, the capacity share ended up at 65 MW, or 0.4%, by the end of the 9th Plan, and biogas contributes 4.95 MW of the grid-connected capacity [67]. Until 2015, the share of fossil fuels within the world energy supply was relatively unchanged, despite the increase of renewable energy in electricity generation to 34% of the global figure [91]. Most of the electricity mix in Malaysia was generated from fossil fuels, although the country has various renewable energy sources.

Apparently, biogas can be one of the best alternatives energy sources to deal with high energy demands and various environmental loads, including fossil fuel depletion and global warming. In addition, biogas adoption can solve waste management issues by utilizing POME for biogas generation. Thus, the environmental sustainability evaluation of biogas production from a broad range of feedstock (e.g., sludge, food waste, dairy manure, municipal wastewater and solid waste, crop residues, energy crops, etc.) is essential in providing a promising renewable energy source. The environmental assessment of biogas would certainly be useful for environmental profile enhancement and a great opportunity to achieve sustainable development. With regard to existing green policy and initiatives, the LCA of biogas production should also be integrated into the sustainable development plan, as this approach could assist in decision-making process.

5. Conclusions

A case study concerning the LCA of the zero waste discharge treatment of POME has been conducted in order to examine the key advances in waste-to-energy technologies that have been adopted for biogas production and waste treatment towards sustainable development. The zero waste discharge treatment system is said to be a promising sustainable technology because it can produce biogas with maximum water recycle and reuse.

In promoting biogas as a green product, Malaysia could gain competitive advantages towards renewable energy production, as well as towards better waste management practices. Therefore, a comprehensive framework enforcement is needed to encourage the embracement of renewable energy and stimulate an energy efficiency culture. A conceptual framework for an LCA-based environmental sustainability evaluation of biogas production has been proposed to improve the sustainability of biogas supply chains. The conceptual framework developed in the present study can be used as a guideline for the related stakeholders and decision makers to improve the quality and enhance the sustainability of biogas supply chains in Malaysia, as well as to promote biogas as a clean, reliable and secure energy.

To conclude, this study indicated that the biogas production and zero discharge treatment of POME have potential as clean technologies to be applied in the Malaysian context, as the system utilized organic waste to produce bioenergy and achieved zero discharge. However, there were other aspects that should be taken into consideration, particularly regarding the sources of electricity and upstream activity, to ensure the environmental sustainability of the system holistically.

Author Contributions: Conceptualization, N.I.H.A.A. and M.M.H.; methodology, N.I.H.A.A.; validation, N.I.H.A.A., M.M.H. and S.H.G.; formal analysis, N.I.H.A.A.; writing—original draft preparation, N.I.H.A.A.; writing—review and editing, M.M.H. and S.H.G.; visualization, H.I.; supervision, M.M.H.; project administration, M.M.H.; funding acquisition, M.M.H. All authors have read and agreed to the published version of the manuscript.

Funding: The APC was funded by the Universiti Kebangsaan Malaysia research grants DIP-2019-001 and MI-2020-005.

Acknowledgments: Marlia M. Hanafiah was supported by the Universiti Kebangsaan Malaysia research grants DIP-2019-001, KRA-2018-054 and MI-2020-005, and by the Ministry of Education of Malaysia (FRGS/1/2018/WAB05/UKM/02/2).

Conflicts of Interest: The authors declare no conflicts of interest.

Abbreviations

GWP	global warming
ODP	stratospheric ozone depletion
IRP	ionizing radiation
HOFP	ozone formation (human health)
PMFP	fine particulate matter formation
EOFP	ozone formation (terrestrial ecosystems)
TAP	terrestrial acidification
FEP	freshwater eutrophication
MEP	marine eutrophication
TETP	terrestrial ecotoxicity
FETP	freshwater ecotoxicity
METP	marine ecotoxicity
HTPc	human carcinogenic toxicity
HTPnc	human non-carcinogenic toxicity
SOP	mineral resource scarcity
FFP	fossil resource scarcity
WCP	water consumption

References

1. Hezri, A.A.; Hasan, M.N. Towards sustainable development? The evolution of environmental policy in Malaysia. *Nat. Resour. Forum* **2006**, *30*, 37–50. [CrossRef]
2. Mohamed, A.R.; Lee, K.T. Energy for sustainable development in Malaysia: Energy policy and alternative energy. *Energy Policy* **2006**, *34*, 2388–2397. [CrossRef]
3. Seman, S.Z.A.; Idris, I.; Abdullah, A.; Samsudin, I.K.; Othman, M.R. Optimizing purity and recovery of biogas methane enrichment process in a closed landfill. *Renew. Energy* **2019**, *131*, 1117–1127. [CrossRef]
4. Griggs, D.; Stafford-Smith, M.; Gaffney, O.; Rockström, J.; Öhman, M.C.; Shyamsundar, P.; Steffen, W.; Glaser, G.; Kanie, N.; Noble, I. Policy: Sustainable development goals for people and planet. *Nature* **2013**, *495*, 305–307. [CrossRef]
5. Hanafiah, M.M.; Mohamed Ali, M.Y.; Abdul Aziz, N.I.H.; Ashraf, M.A.; Halim, A.A.; Lee, K.E.; Idris, M. Biogas production from goat and chicken manure in Malaysia. *Appl. Ecol. Environ. Res.* **2017**, *15*, 529–535. [CrossRef]
6. Nagy, V.; Szabó, E. Biogas from organic wastes. *Studia Univ. Vasile Goldis Arad Ser. Stiintele Vietii Life Sci. Ser.* **2011**, *21*, 887–891.
7. Aziz, N.I.H.A.; Hanafiah, M.M. Anaerobic digestion of palm oil mill effluent (POME) using bio-methane potential (BMP) test. *AIP Conf. Proc.* **2018**, *1940*, 020026.
8. Horváth, I.S.; Tabatabaei, M.; Karimi, K.; Kumar, R. Recent updates on biogas production—A review. *Biofuel Res. J.* **2016**, *3*, 394–402. [CrossRef]
9. Aziz, N.I.H.A.; Hanafiah, M.M. The potential of palm oil mill effluent (POME) as a renewable energy source. *J. Green Energy* **2017**, *1*, 323–346. [CrossRef]
10. Bujang, A.; Bern, C.; Brumm, T. Summary of energy demand and renewable energy policies in Malaysia. *Renew. Sustain. Energy Rev.* **2016**, *53*, 1459–1467. [CrossRef]
11. Roubík, H.; Mazancová, J. Small-scale biogas plants in central Vietnam and biogas appliances with a focus on a flue gas analysis of biogas cook stove. *Renew. Energy* **2019**, *131*, 1138–1145. [CrossRef]
12. Scarlat, N.; Dallemand, J.-F.; Fahl, F. Biogas: Developments and perspectives in Europe. *Renew. Energy* **2018**, *129*, 457–472. [CrossRef]
13. Eriksson, O.; Bisaillon, M.; Haraldsson, M.; Sundberg, J. Enhancement of biogas production from food waste and sewage sludge–environmental and economic life cycle performance. *J. Environ. Manag.* **2016**, *175*, 33–39. [CrossRef]
14. Yasar, A.; Rasheed, R.; Tabinda, A.B.; Tahir, A.; Sarwar, F. Life cycle assessment of a medium commercial scale biogas plant and nutritional assessment of effluent slurry. *Renew. Sustain. Energy Rev.* **2017**, *67*, 364–371. [CrossRef]
15. Finnveden, G.; Hauschild, M.Z.; Ekvall, T.; Guinée, J.; Heijungs, R.; Hellweg, S.; Koehler, A.; Pennington, D.; Suh, S. Recent developments in life cycle assessment. *J. Environ. Manag.* **2009**, *91*, 1–21. [CrossRef]
16. Hanafiah, M.M. Quantifying Effects of Physical, Chemical and Biological Stressors in Life Cycle Assessment. Ph.D Thesis, Radboud University, Nijmegen, The Netherlands, 2013.
17. Rebitzer, G.; Ekvall, T.; Frischknecht, R.; Hunkeler, D.; Norris, G.; Rydberg, T.; Schmidt, W.P.; Suh, S.; Weidema, B.P.; Pennington, D.W. Life cycle assessment: Part 1: Framework, goal and scope definition, inventory analysis and applications. *Environ. Int.* **2004**, *30*, 701–720. [CrossRef]
18. Curran, M.A. *Life Cycle Assessment: Principles and Practice*; United States Environmental Protection Agency: Washington, DC, USA, 2006.
19. Heijungs, R.; Goedkoop, M.; Struijs, J.; Effting, S.; Sevenster, M.; Huppes, G. Towards a Life Cycle Impact Assessment Method Which Comprises Category Indicators at the Midpoint and the Endpoint Level. Report of the First Project Phase: Design of the New Method VROM Report. 2003. Available online: http://www.leidenuniv.nl/cml/ssp/publications/recipe_phase1.pdf (accessed on 1 March 2019).
20. Huijbregts, M. A critical view on scientific consensus building in life cycle impact assessment. *Int. J. Life Cycle Assess.* **2014**, *19*, 477–479. [CrossRef]
21. Talve, S. The life cycle assessment management tool for technologies in Eastern Europe: Why and how. *Int. J. Life Cycle Assess.* **2001**, *6*, 181–183. [CrossRef]
22. Hosseini, S.E.; Wahid, M.A. Feasibility study of biogas production and utilization as a source of renewable energy in Malaysia. *Renew. Sustain. Energy Rev.* **2013**, *19*, 454–462. [CrossRef]

23. Ali, M.Y.M.; Hanafiah, M.M.; Wen, Y.H.; Idris, M.; Aziz, N.I.H.A.; Halim, A.A.; Lee, K.E. Biogas production from different substrates under anaerobic conditions. In Proceedings of the 3rd International Conference on Agricultural and Medical Sciences (CAMS-2015), Singapore, 10–11 December 2015; pp. 54–56.
24. Aziz, N.I.H.A.; Hanafiah, M.M.; Ali, M.Y.M. Sustainable biogas production from agrowaste and effluents–a promising step for small-scale industry income. *Renew. Energy* **2019**, *132*, 363–369. [CrossRef]
25. Choo, Y.M.; Muhamad, H.; Hashim, Z.; Subramaniam, V.; Puah, C.W.; Tan, Y. Determination of GHG contributions by subsystems in the oil palm supply chain using the LCA approach. *Int. J. Life Cycle Assess.* **2011**, *16*, 669–681. [CrossRef]
26. SIRIM. LCA Projects. 2016. Available online: http://lcamalaysia.sirim.my/index.php/project (accessed on 5 March 2019).
27. Aziz, N.I.H.A.; Hanafiah, M.M. Life cycle analysis of biogas production from anaerobic digestion of palm oil mill effluent. *Renew. Energy* **2020**, *145*, 847–857. [CrossRef]
28. Lamnatou, C.; Nicolai, R.; Chemisana, D.; Cristofari, C.; Cancelleiri, D. Biogas production means of an anaerobic-digestion plant in France: LCA of greenhouse-gas emissions and other environmental indicators. *Sci. Total Environ.* **2019**, *670*, 1226–1239. [CrossRef]
29. Lauer, M.; Dotzauer, M.; Hennig, C.; Lehmann, M.; Nebel, E.; Postel, J.; Szarka, N.; Thrän, D. Flexible power generation scenarios for biogas plants operated in Germany: Impacts on economic viability and GHG emissions. *Int. J. Energy Res.* **2017**, *41*, 63–80. [CrossRef]
30. Guinée, J.B. Handbook on life cycle assessment operational guide to the ISO standards. *Int. J. Life Cycle Assess.* **2002**, *7*, 311–313. [CrossRef]
31. Hussein, M.E.; Alam, R.Z.; Siwar, C.; Ludin, N.A. Green economy models and energy policies towards sustainable development in Malaysia: A review. *Int. J. Green Econ.* **2016**, *10*, 89–106. [CrossRef]
32. Yatim, P.; Mamat, M.N.; Mohamad Zailani, S.H.; Ramlee, S. Energy policy shifts towards sustainable energy future for Malaysia. Clean Technol. *Environ. Policy* **2016**, *18*, 1685–1695.
33. Chen, W.M.; Kim, H.; Yamaguchi, H. Renewable energy in eastern Asia: Renewable energy policy review and comparative SWOT analysis for promoting renewable energy in Japan, South Korea and Taiwan. *Energy Policy* **2014**, *74*, 319–329. [CrossRef]
34. Hua, Y.; Oliphant, M.; Hu, E.J. Development of renewable energy in Australia and China: A comparison of policies and status. *Renew. Energy* **2016**, *85*, 1044–1051. [CrossRef]
35. International Energy Agency (IEA). Policy and Measures, Renewable Energy. 2019. Available online: www.iea.org (accessed on 17 March 2019).
36. Alam, S.S.; Nor, N.F.M.; Ahmad, M.; Hashim, N.H.N. A survey on renewable energy development in Malaysia: Current status, problems and prospects. *Environ. Climate Technol.* **2016**, *17*, 5–17. [CrossRef]
37. Fang, T.P.; Daud, W.R.W.; Halim, L.; Masdar, M.S. How ready is renewable energy? A review on renewable energy and fuel cell teaching in schools. In Proceedings of the 7th World Engineering Education Forum (WEEF), Kuala Lumpur, Malaysia, 13–16 November 2017; pp. 236–244.
38. Brundtland, G. *Our Common Future;* Brundtland Report; United Nations World Commission on Environment and Development: Geneva, Switzerland, 1987.
39. Heijungs, R.; Huppes, G.; Guinée, J.B. Life cycle assessment and sustainability analysis of products, materials and technologies. Toward a scientific framework for sustainability life cycle analysis. *Polym. Degrad. Stab.* **2010**, *95*, 422–428. [CrossRef]
40. Kloepffer, W. Life cycle sustainability assessment of products. *Int. J. Life Cycle Assess.* **2008**, *13*, 89. [CrossRef]
41. Mälkki, H.; Alanne, K. An overview of life cycle assessment (LCA) and researched-based teaching in renewable and sustainable energy education. *Renew. Sustain. Energy Rev.* **2017**, *69*, 218–231. [CrossRef]
42. Coyle, E.; Rebow, M. Sustainable Design: A Case Study in Energy Systems. In *Engineering in Context;* Dublin Institute of Technology: Dublin, Ireland, 2009; pp. 1–17.
43. Jürgensen, L.; Ehimen, E.A.; Born, J.; Holm-Nielsen, J.B. A combination anaerobic digestion scheme for biogas production from dairy effluent—CSTR and ABR, and biogas upgrading. *Biomass Bioenergy* **2018**, *111*, 241–247. [CrossRef]
44. Ertem, F.C.; Martínez-Blanco, J.; Finkbeiner, M.; Neubauer, P.; Junne, S. Life cycle assessment of flexibly fed biogas processes for an improved demand-oriented biogas supply. *Bioresour. Technol.* **2016**, *219*, 536–544. [CrossRef]

45. Styles, D.; Dominguez, E.M.; Chadwick, D. Environmental balance of the of the UK biogas sector: An evaluation by consequential life cycle assessment. *Sci. Total Environ.* **2016**, *560*, 241–253. [CrossRef]
46. Curran, M. *A Review of Life-Cycle Based Tools Used to Assess the Environmental Sustainability of Biofuels in the United States*; Life Cycle Assessment Research Center; US Environmental Protection Agency: Cincinnati, OH, USA, 2013; Volume 1, p. 61.
47. Sala, S.; Farioli, F.; Zamagni, A. Progress in sustainability science: Lessons learnt from current methodologies for sustainability assessment: Part 1. *Int. J. Life Cycle Assess.* **2013**, *18*, 1653–1672. [CrossRef]
48. Takeda, S.; Keeley, A.R.; Sakurai, S.; Managi, S.; Norris, C.B. Are renewables as friendly to humans as to the environment? A social life cycle assessment of renewable electricity. *Sustainability* **2019**, *11*, 1370. [CrossRef]
49. Markevičius, A.; Katinas, V.; Perednis, E.; Tamašauskienė, M. Trends and sustainability criteria of the production and use of liquid biofuels. *Renew. Sustain. Energy Rev.* **2010**, *14*, 3226–3231. [CrossRef]
50. Milazzo, M.; Spina, F.; Cavallaro, S.; Bart, J. Sustainable soy biodiesel. *Renew. Sustain. Energy Rev.* **2013**, *27*, 806–852. [CrossRef]
51. Turconi, R.; Boldrin, A.; Astrup, T. Life cycle assessment (LCA) of electricity generation technologies: Overview, comparability and limitations. *Renew. Sustain. Energy Rev.* **2013**, *28*, 555–565. [CrossRef]
52. Bößner, S.; Devisscher, T.; Suljada, T.; Ismail, C.J.; Sari, A.; Mondamina, N.W. Barriers and opportunities to bioenergy transitions: An integrated, multi-level perspective analysis of biogas uptake in Bali. *Biomass Bioenergy* **2019**, *122*, 457–465. [CrossRef]
53. Meyer, A.K.P.; Ehimen, E.A.; Holm-Nielsen, J.B. Future European biogas: Animal manure, straw and grass potentials for a sustainable European biogas production. *Biomass Bioenergy* **2018**, *111*, 154–164. [CrossRef]
54. Basiron, Y. Palm oil production through sustainable plantations. *Eur. J. Lipid Sci. Technol.* **2007**, *109*, 289–295. [CrossRef]
55. Ming, T.C.; Ramli, N.; Lye, O.T.; Said, M.; Kasim, Z. Strategies for decreasing the pour point and cloud point of palm oil products. *Eur. J. Lipid Sci. Technol.* **2005**, *107*, 505–512. [CrossRef]
56. Mukherjee, I.; Sovacool, B.K. Palm oil-based biofuels and sustainability in Southeast Asia: A review of Indonesia, Malaysia, and Thailand. *Renew. Sustain. Energy Rev.* **2014**, *37*, 1–12. [CrossRef]
57. Tan, C.; Man, Y.C. Differential scanning calorimetric analysis of palm oil, palm oil based products and coconut oil: Effects of scanning rate variation. *Food Chem.* **2002**, *76*, 89–102. [CrossRef]
58. Teh, S.S.; Ong, A.S.H.; Mah, S.H. Recovery and utilization of palm oil mill effluent source as value-added food products. *J. Oleo Sci.* **2017**, *66*, 1183–1191. [CrossRef]
59. Awalludin, M.F.; Sulaiman, O.; Hashim, R.; Nadhari, W.N.A.W. An overview of the oil palm industry in Malaysia and its waste utilization through thermochemical conversion, specifically via liquefaction. *Renew. Sustain. Energy Rev.* **2015**, *50*, 1469–1484. [CrossRef]
60. Bazmi, A.A.; Zahedi, G.; Hashim, H. Progress and challenges in utilization of palm oil biomass as fuel for decentralized electricity generation. *Renew. Sustain. Energy Rev.* **2011**, *15*, 574–583. [CrossRef]
61. Basri, M.F.; Yacob, S.; Hassan, M.A.; Shirai, Y.; Wakisaka, M.; Zakaria, M.R.; Phang, L.Y. Improved biogas production from palm oil mill effluent by a scaled-down anaerobic treatment process. *World J. Microbiol. Biotechnol.* **2010**, *26*, 505–514. [CrossRef]
62. Oswal, N.; Sarma, P.; Zinjarde, S.; Pant, A. Palm oil mill effluent treatment by a tropical marine yeast. *Bioresour. Technol.* **2002**, *85*, 35–37. [CrossRef]
63. Mekhilef, S.; Barimani, M.; Safari, A.; Salam, Z. Malaysia's renewable energy policies and programs with green aspects. *Renew. Sustain. Energy Rev.* **2014**, *40*, 497–504. [CrossRef]
64. Shafie, S.M.; Mahlia, T.M.I.; Masjuki, H.H.; Andriyana, A. Current energy usage and sustainable energy in Malaysia: A review. *Renew. Sustain. Energy Rev.* **2011**, *15*, 4370–4377. [CrossRef]
65. Sovacool, B.K.; Drupady, I.M. Examining the small renewable energy power (SREP) program in Malaysia. *Energy Policy* **2011**, *39*, 7244–7256. [CrossRef]
66. Umar, M.S.; Jennings, P.; Urmee, T. Strengthening the palm oil biomass renewable energy industry in Malaysia. *Renew. Energy* **2013**, *60*, 107–115. [CrossRef]
67. Umar, M.S.; Jennings, P.; Urmee, T. Sustainable electricity generation from oil palm biomass wastes in Malaysia: An industry survey. *Energy* **2014**, *67*, 496–505. [CrossRef]
68. Haris, A.H. Renewable Energy Bill and Subsidiary Legislations, Ministry of Energy, Green Technology and Water. 2011. Available online: Seda.gov.my (accessed on 13 March 2019).

69. Aziz, N.I.H.A.; Hanafiah, M.M.; Gheewala, S.H. A review on life cycle assessment of biogas production: Challenges and future perspectives in Malaysia. *Biomass Bioenergy* **2019**, *122*, 361–374. [CrossRef]
70. Abdurahman, N.; Rosli, Y.; Azhari, N. The performance evaluation of anaerobic methods for palm oil mill effluent (POME) treatment: A review. In *International Perspectives on Water Quality Management and Pollutant Control*; Quinn, N.W.T., Ed.; IntechOpen: London, UK, 2013; pp. 87–106.
71. Ahmed, Y.; Yaakob, Z.; Akhtar, P.; Sopian, K. Production of biogas and performance evaluation of existing treatment processes in palm oil mill effluent (POME). *Renew. Sustain. Energy Rev.* **2015**, *42*, 1260–1278. [CrossRef]
72. Wu, T.Y.; Mohammad, A.W.; Jahim, J.M.; Anuar, N. Pollution control technologies for the treatment of palm oil mill effluent (POME) through end-of-pipe processes. *J. Environ. Manag.* **2010**, *91*, 1467–1490. [CrossRef]
73. Yuniarto, A.; Noor, Z.Z.; Ujang, Z.; Olsson, G.; Aris, A.; Hadibarata, T. Bio-fouling reducers for improving the performance of an aerobic submerged membrane bioreactor treating palm oil mill effluent. *Desalination* **2013**, *316*, 146–153. [CrossRef]
74. Durham, B.; Mierzejewski, M. Water reuse and zero liquid discharge: A sustainable water resource solution. *Water Sci Tech. Water Supply* **2003**, *3*, 97–103. [CrossRef]
75. Heins, W.; Schooley, K. Achieving zero liquid discharge in SAGD heavy oil recovery. *J. Can. Petrol. Technol.* **2004**, *43*, 1–8. [CrossRef]
76. Katsoyiannis, I.; Castellana, M.; Cartechini, F.; Vaccarella, A.; Zouboulis, A.; Grinias, K. Application of zero liquid discharge water treatment units for wastewater reclamation: Possible application in marine ports. In *Sustainable Development of Sea-Corridors and Coastal Waters*; Springer: Cham, Switzerland, 2015; pp. 39–45.
77. Mu, C.; Lin, W.; Zhang, M.; Zhu, Q. Towards zero discharge of chromium-containing leather waste through improved alkali hydrolysis. *Waste Manag.* **2003**, *23*, 835–843. [CrossRef]
78. Wang, J.; Mahmood, Q.; Qiu, J.P.; Li, Y.S.; Chang, Y.S.; Chi, L.N.; Li, X.D. Zero discharge performance of an industrial pilot-scale plant treating palm oil mill effluent. *BioMed Res. Int.* **2015**, *2015*, 1–9. [CrossRef]
79. Rajakumari, S.P.; Kanmani, S. Environmental life cycle assessment of zero liquid discharge treatment technologies for textile industries, Tirupur: A case study. *J. Sci. Ind. Res.* **2008**, *67*, 461–467.
80. Grant, S.B.; Saphores, J.D.; Feldman, D.L.; Hamilton, A.J.; Fletcher, T.D.; Cook, P.L.; Stewardson, M.; Sanders, B.F.; Levin, L.A.; Ambrose, R.F. Taking the "waste" out of "wastewater" for human water security and ecosystem sustainability. *Science* **2012**, *337*, 681–686. [CrossRef]
81. Madaki, Y.S.; Seng, L. Pollution control: How feasible is zero discharge concepts in Malaysia palm oil mills. *Am. J. Engineer. Res.* **2013**, *2*, 239–252.
82. Tong, T.; Elimelech, M. The global rise of zero liquid discharge for wastewater management: Drivers, technologies and future directions. *Environ. Sci. Technol.* **2016**, *50*, 6846–6855. [CrossRef]
83. Weidema, B.P.; Bauer, C.; Hischier, R.; Mutel, C.; Nemecek, T.; Reinhard, J.; Vadenbo, C.; Wernet, G. *Overview and Methodology: Data Quality Guideline for the Ecoinvent Database Version 3*; Ecoinvent Report 1 (v3); The Ecoinvent Centre: St. Gallen, Switzerland, 2013.
84. Durlinger, B.; Tyszler, M.; Scholten, J.; Broekema, R.; Blonk, H.; Beatrixstraat, G. Agri-footprint: A life cycle inventory database covering food and feed production and processing. In Proceedings of the 9th International Conference on Life Cycle Assessment in the Agri-Food Sector, San Francisco, CA, USA, 8–10 October 2014; pp. 310–317.
85. Norris, G.A. *Simapro Database Manual: The Franklin US LCI Library*; Pré Consultants and Sylvatica: Amersfoort, The Netherlands, 2004; pp. 1–30.
86. Huijbregts, M.; Steinmann, Z.; Elshout, P.; Stam, G.; Verones, F.; Vieira, M.; Hollander, A.; Zijp, M.; Van Zelm, R. *Recipe 2016: A Harmonized Life Cycle Impact Assessment Method at Midpoint and Endpoint Level Report I: Characterization*; RIVM Report 2016–0104; National Institute for Human Health and the Environment: Bilthoven, The Netherlands, 2016.
87. ISO 14044. *Environmental Management—Life Cycle Assessment—Requirements and Guidelines, European Committee for Standardization*; International Organization for Standardization ISO Central Secretariat: Geneva, Switzerland, 2006.
88. Lee, K.E.; Mokhtar, M.; Mohd Hanafiah, M.; Abdul Halim, A.; Badusah, J. Rainwater harvesting as an alternative water resource in Malaysia: Potential, policies and development. *J. Clean. Prod.* **2016**, *126*, 218–222. [CrossRef]

89. Loh, S.K.; Lai, M.E.; Ngatiman, M.; Lim, W.S.; Choo, Y.M.; Zhang, Z.; Salimon, J. Zero discharge treatment technology of palm oil mill effluent. *J. Oil Palm Res.* **2013**, *25*, 273–281.
90. Tabassum, S.; Zhang, Y.; Zhang, Z. An integrated method for palm oil mill effluent (POME) treatment for achieving zero liquid discharge: A pilot study. *J. Clean. Prod.* **2015**, *95*, 148–155. [CrossRef]
91. International Energy Agency. *CO_2 Emissions from Fuel Combustion—Highlights*; International Energy Agency: Paris, France, 2017.
92. Energy Commission. 2016. Available online: https://www.st.gov.my/ (accessed on 3 January 2019).
93. Oh, T.H.; Hasanuzzaman, M.; Selvaraj, J.; Teo, S.C.; Chua, S.C. Energy policy and alternative energy in Malaysia: Issues and challenges for sustainable growth–an update. *Renew. Sustain. Energy Rev.* **2017**, *81*, 3021–3031. [CrossRef]
94. Mahlia, T. Emissions from electricity generation in Malaysia. *Renew. Energy* **2002**, *27*, 293–300. [CrossRef]
95. Shafie, S.; Mahlia, T.; Masjuki, H.; Ahmad-Yazid, A. A review on electricity generation based on biomass residue in Malaysia. *Renew. Sustain. Energy Rev.* **2012**, *16*, 5879–5889. [CrossRef]
96. Lijó, L.; González-García, S.; Bacenetti, J.; Fiala, M.; Feijoo, G.; Lema, J.M.; Moreira, M.T. Life cycle assessment of electricity production in Italy from anaerobic co-digestion of pig slurry and energy crops. *Renew. Energy* **2014**, *68*, 625–635. [CrossRef]
97. Wang, Q.L.; Li, W.; Gao, X.; Li, S.J. Life cycle assessment on biogas production from straw and its sensitivity analysis. *Bioresour. Technol.* **2016**, *201*, 208–214. [CrossRef]
98. Roux, C.; Schalbart, P.; Peuportier, B. Accounting for temporal variation of electricity production and consumption in the LCA of an energy-efficient house. *J. Clean. Prod.* **2016**, *113*, 532–540. [CrossRef]
99. Wang, J.; Mahmood, Q.; Qiu, J.P.; Li, Y.S.; Chang, Y.S.; Li, X.D. Anaerobic treatment of palm oil mill effluent in pilot-scale anaerobic EGSB reactor. *BioMed Res. Int.* **2015**, *2015*, 1–7. [CrossRef]
100. Ali, R.; Daut, I.; Taib, S. A review on existing and future energy sources for electrical power generation in Malaysia. *Renew. Sustain. Energy Rev.* **2012**, *16*, 4047–4055. [CrossRef]
101. Khor, C.S.; Lalchand, G. A review on sustainable power generation in Malaysia to 2030: Historical perspective, current assessment, and future strategies. *Renew. Sustain. Energy Rev.* **2014**, *29*, 952–960. [CrossRef]
102. Jaramillo, P.; Griffin, W.M.; Matthews, H.S. Comparative life-cycle air emissions of coal, domestic natural gas, lng, and sng for electricity generation. *Environ. Sci. Technol.* **2007**, *41*, 6290–6296. [CrossRef] [PubMed]
103. Kumar, S.; Kwon, H.T.; Choi, K.H.; Hyun Cho, J.; Lim, W.; Moon, I. Current status and future projections of LNG demand and supplies: A global prospective. *Energy Policy* **2011**, *39*, 4097–4104. [CrossRef]
104. Tilton, J.E. *On Borrowed Time? Assessing the Threat of Mineral Depletion*; Routledge: New York, NY, USA, 2010; pp. 1–158.
105. International Gas Union. 2017. Available online: https://www.igu.org/ (accessed on 3 January 2019).
106. Hengeveld, E.J.; Bekkering, J.; Van Gemert, W.J.T.; Broekhuis, A.A. Biogas infrastructures from farm to regional scale, prospects of biogas transport grids. *Biomass Bioenergy* **2016**, *86*, 43–52. [CrossRef]
107. Trummer, D.R. Biomass-Fired CHP in Palm Oil Mills. In *Malaysian-Danish Country Programme for Cooperation in Environment and Sustainable Development (2002–2006)*; Royal Danish Embassy: Kuala Lumpur, Malaysia, 2001.
108. Cervi, W.R.; Lamparelli, R.A.C.; Seabra, J.E.A.; Junginger, M.; Van Der Hilst, F. Bioelectricity potential from ecologically available sugarcane straw in Brazil: A spatially explicit assessment. *Biomass Bioenergy* **2019**, *122*, 391–399. [CrossRef]
109. Gigliucci, G.; Petruzzi, L.; Cerelli, E.; Garzisi, A.; La Mendola, A. Demonstration of a residential CHP system based on PEM fuel cells. *J. Power Sources* **2004**, *131*, 62–68. [CrossRef]
110. Parsaee, M.; Kiani, M.K.D.; Karimi, K. A review of biogas production from sugarcane vinasse. *Biomass Bioenergy* **2019**, *122*, 117–125. [CrossRef]
111. Pöschl, M.; Ward, S.; Owende, P. Evaluation of energy efficiency of various biogas production and utilization pathways. *Appl. Energy* **2010**, *87*, 3305–3321. [CrossRef]
112. Chin, M.J.; Poh, P.E.; Tey, B.T.; Chan, E.S.; Chin, K.L. Biogas from palm oil mill effluent (POME): Opportunities and challenges from Malaysia's perspective. *Renew. Sustain. Energy Rev.* **2013**, *26*, 717–726. [CrossRef]

sustainability

MDPI

Review

Recycling Organic Fraction of Municipal Solid Waste: Systematic Literature Review and Bibliometric Analysis of Research Trends

José María Fernández-González [1], Carmen Díaz-López [2], Jaime Martín-Pascual [2] and Montserrat Zamorano [2,*]

[1] PROMA, Proyectos de Ingeniería Ambiental, S.L. Gran Vía, 48, 1810 Granada, Spain; josemaria@promaingenieros.com

[2] Department of Civil Engineering, ETS Ingeniería de Caminos, Canales y Puertos, University of Granada, 18071 Granada, Spain; carmendiaz@ugr.es (C.D.-L.); jmpascual@ugr.es (J.M.-P.)

* Correspondence: zamorano@ugr.es; Tel.: +34-958-249-458

Received: 13 May 2020; Accepted: 3 June 2020; Published: 11 June 2020

Abstract: The organic fraction is usually the predominant fraction in municipal solid waste, so its recycling is a potential alternative to disposal in landfill sites, as well as helping to reach targets included in the European Circular Economy Package. The existing body of knowledge in this research field is very large, so a comprehensive review of the existing scientific literature has been considered of interest to provide researchers and professionals with a detailed understanding of the status quo and predict the dynamic directions of this field. A systematic literature review and bibliometric analysis have been performed to provide objective criteria for evaluating the work carried out by researchers and a macroscopic overview of the existing body of knowledge in this field. The analysis of 452 scientific articles published from 1980 to 2019 has shown that the application of composting technologies is relevant, especially since 2014, when policies aimed at reducing emissions to the atmosphere were increased and focused on the use of this waste fraction to produce biogas. Nevertheless, the scientific field is still evolving to impose a model of a circular economy; in fact, emerging studies are being conducted on the production of biomethane, contributing to the decarbonised energy system.

Keywords: bibliometric analysis; analysis of science mapping; SciMAT; systematic literature review; municipal solid waste; organic fraction

---

1. Introduction

Municipal solid waste (MSW) management is an important challenge of the urban environment in most cities worldwide today. In fact, a big problem concerns the planning of treatment plants that can face the quantity and composition of municipal waste [1]; as a consequence, sustainable management solutions should be designed [2]. In this respect, several studies have reported that the composition of MSW varies significantly from one country, region or municipality to another, including food waste, metals, plastics, glass, textiles or inert materials, among others [3]. The composition of MSW depends significantly on factors such as lifestyle, economic level or legal framework, and knowledge of it is critical to determining the appropriate handling and management of these wastes [4,5]. However, the organic fraction (OFMSW) is usually the predominant fraction, in the case of its selective collection not being implemented; in fact, the percentage of the OFMSW worldwide is 46%, varying between 64% in developing countries and 28% in the case of higher-income ones [6]. Although the data indicate a slight reduction in the proportion of organic waste in 2025, it is forecast that solid waste generation rates will exceed 11 million metric tons per day, which is more than three times the current rate, for the year of 2100 [6]; given that the amount of organic waste will increase along with the total amount of solid waste [7], its recycling has attracted a great deal of attention as a potential alternative to the conventional solid waste disposal of a wide range of residues in landfill sites, mainly for the countries outside of the EU, where waste regulation is not so stringent.

Traditionally, the most commonly used technologies for the recycling of the OFMSW are composting and anaerobic digestion to produce compost and biogas and digestate, respectively [8–10]. Compost and digestate are rich in nutrients, so they are used as organic amendments in soils [11]; the methane content in biogas makes it a source of renewable energy [12]. These treatments are widely known and applied today; however, the search for systems that reduce some of the limits of aerobic digestion (for example, the problems of odours, occupation of space and degradation time) [7,13] or improve performance in terms of the production of methane [14,15] have led to technological development, for example, the co-composting of OFMSW and another organic solid waste such as faecal sludge [16,17] and animal and/or agricultural waste [18–20], among others, or the reaction of wet organic substrates under hydrothermal conditions or hydrothermal carbonisation (HTC) [21]

On the other hand, increasingly tight regulations, for example, Directive (EU) 2018/850 amending Directive 1999/31/EC on landfill waste, as well as the increasing demand for renewable fuels, are driving the conversion of the wastes into valuable bio-products that can substitute non-renewable materials and ensure the effective use of existing resources through circular flow loops in a sustainable way. In fact, the European Commission has recently adopted an ambitious Circular Economy Package, which includes measures to drive Europe's transition towards this economic model, which includes a common EU target for recycling 65% of municipal waste and reducing landfill to a maximum of 10% of municipal waste by 2030 [22]. To reach these targets, recycling technologies applied to the organic matter of municipal solid waste now represent an important strategy in the area of waste treatment; this may be perceived as a potential alternative to provide a renewable source of energy, as well as to use the recycling potential of the biodegradable fraction of waste generated by a large number of activities [23].

On account of the important current research to promote the recycling of OFMSW as well as the great diversity of applicable technologies, uses of by-products, construction and operation costs, social acceptance, environmental impacts or measures that promote the implementation of treatment systems for this fraction, it is difficult to obtain a single point from which to access this topic. This diversity also leads to a lack of a broad view of the area of research or the evolution of issues in this field, which makes it difficult to obtain useful and impartial information for future work. Therefore, developing a comprehensive review of the existing scientific literature has been considered of interest to make it easier to integrate the contributions in order to obtain a critical perspective.

In this sense, the bibliometric analysis provides objective criteria to evaluate the research carried out in a field [24], as well as a macroscopic overview of large amounts of scientific literature [25]. However bibliographic study dates back in a particular field to the 19th century; Alan Pritchard coined the concept of bibliometric analysis in 1969 [25]. This methodology has grown exponentially with the development of the internet, which has made easier and faster the communication between researchers, as well as the access to contributions in a given field around the world [26].

Performance analysis and science mapping are the two main methods of bibliometric research. The first one aims at the evaluation of the impact of citations in scientific production; on the other hand, science mapping defines the conceptual, social and intellectual scientific research structure, and its evolution. These methods show a representation of the relationships between the disciplines, fields, specialities, documents or authors [27]; they also examine the bibliographic material from an objective and quantitative perspective [28]. In consequence, many disciplines use these methods to study the impact of their field, researchers, or a particular document, in order to determine the structural and dynamic characteristics of scientific research [29].

The objective of this study was thus to develop a bibliometric analysis of the organic fraction of municipal solid waste recycling treatments, through the performance of (i) a systematic review for a quantitative analysis, (ii) a qualitative review using a science mapping study and (iii) the analysis of results.

This study will establish research themes, mapping researcher networks and recommendations for future studies in the research field of recycling OFMSW, contributing to the existing body of knowledge by assessing and highlighting the patterns and trends in the research field.

2. Materials and Methods

Figure 1 shows the double integrated analysis performed to achieve the objectives of this study, including: (i) a systematic literature review (SLR) of the bibliographic records on recycling the OFMSW, and (ii) a bibliometric analysis of the identified documents. The sections below describe each of these procedures.

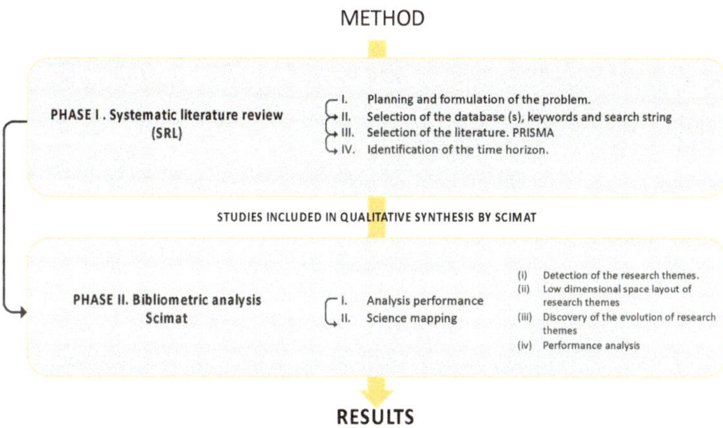

Figure 1. Materials and methods.

2.1. Systematic Literature Review

The SLR establishes a methodological approach that compiles all the empirical evidence that conforms to pre-specified eligibility criteria [30], in order to establish useful findings in the literature [31]. The SLR identifies any gap by minimising research bias and providing reliable results from which conclusions can be drawn and decisions made [32]. The review has been carried out following a search strategy and presents evidence on data sources, selection criteria and analysis [33] carried out according to the following four stages (Figure 1), according to Kitchenham's guidelines [34]:

- *Planning and formulation of the problem.* In this step, the SLR is planned, the problem is formulated and the scope of the review is set. To do that, it is necessary to have a clear definition of the proposed research questions, the exclusion criteria for the final selection of the significant documents and the definition of the expected results.
- *Selection of the database (s), keywords and the search string.* The second step defines the most appropriate bibliographic databases, search string and keywords for the document search. The main problem in carrying out searches in the databases is to determine the keywords and search chains that allow scientific documents relevant to the objectives of the research to be identified. In consequence, it should be necessary to have a large enough number of keywords to not restrict the number of studies, but it should be also specific enough to include only studies related to the research field under study. The first set of pre-selected records could be obtained thanks to the application of a search string.
- *Selection of the literature.* This is a key step to guarantee the selection of a significant number of relevant documents. The most relevant documents are those enclosing the data necessary to

address the research questions of the SLR. They will be selected following the guidelines of the PRISMA flow diagram [35]. Flow diagram is a collective term for a diagram representing a flow or set of dynamic relationships in a system.

- *Identification of time horizon; selection of the database(s).* Finally, and before science mapping, it is essential to establish different periods based on the number of relevant documents identified, as well as the main elements and inflection points of the research field.

2.2. Bibliometric Analysis: Performance Analysis and Science Mapping

Bibliometric analysis was carried out by performance analysis and science mapping; its objective was the obtainment of a spatial illustration of the connection between disciplines, specialities, individual documents and authors [28]. The performance analysis quantifies the impact of the citation of scientific production; on the other hand, the mapping of science shows the social analysis, and the intellectual and conceptual evolution in field research, as well as its evolution and dynamic characteristics. To do that, the free scientific mapping tool SciMAT (Science Mapping Analysis Software Tool) [28] was used. This software is based on the analysis of co-words and the h-Index, incorporating methods, algorithms and measurements in the workflow of the general science mapping, from preprocessing to the visualisation of results [28].

The application of SciMAT enables the detection of the research themes, where an equivalence index [28] is generated, followed by the clustering of thematic keywords using the simple centres algorithm [36] in order to identify the most relevant themes. It continues with the creation of two-dimensional strategic diagrams based on the degree of interaction of different research topics (centrality) and the internal strength value of the research topic object of study (density). In these diagrams, the following four different research topics are reflected by periods (Figure 2a):

- Motor themes. These include important and developed topics in the research field.
- Highly developed and isolated themes. These are well developed topics, but, unlike motor themes, they are not important for the research field object of study.
- Emerging and declining themes. These include little-developed and non-important topics in the research field.
- Basic and transversal themes. These are important topics in the research field object of study, but they are not well developed.

Subsequently, conceptual links between research topics, in different periods, as well as the strength of association between the themes through the inclusion index [37] are detected. The following two types of graphics are used for their representation:

- Overlay graph (Figure 2b). The number of words shared by both periods is represented on the horizontal arrow. The upper incoming arrow shows the number of new words in period 2, and the upper outgoing one shows the words that disappear in period 2.
- Thematic evolution map (Figure 2c). The solid lines mean that the linked theme shares the main item; on the other hand, a dotted line means that the themes share elements that are not the main item. The volumes of the spheres are proportional to the numbers of published documents, and the thicknesses of the edges are proportional to the Inclusion Indices.

Additionally, the contribution of research topics to the whole field of research is quantitatively and qualitatively measured by bibliometric measurements. Among others, the number of documents published, number of citations of the documents, most cited authors, most cited publications and different variants of the h-index [38] are detected.

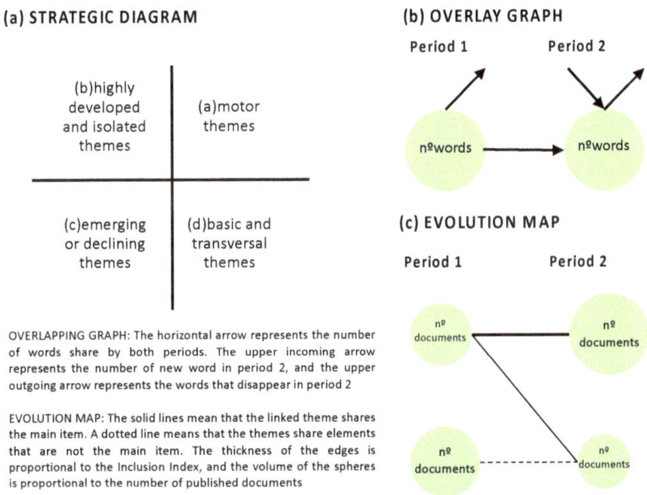

Figure 2. Example of (**a**) a strategic diagram, (**b**) overlay graph and (**c**) evolution map.

3. Results and Discussion

The SLR method and the science mapping study of the relevant documents were applied to carry out an exhaustive analysis of the OFMSW recycling treatment research field. The results obtained are summarised in Figures 3–6 and Tables 1–4, and they are described in detail in the following sections.

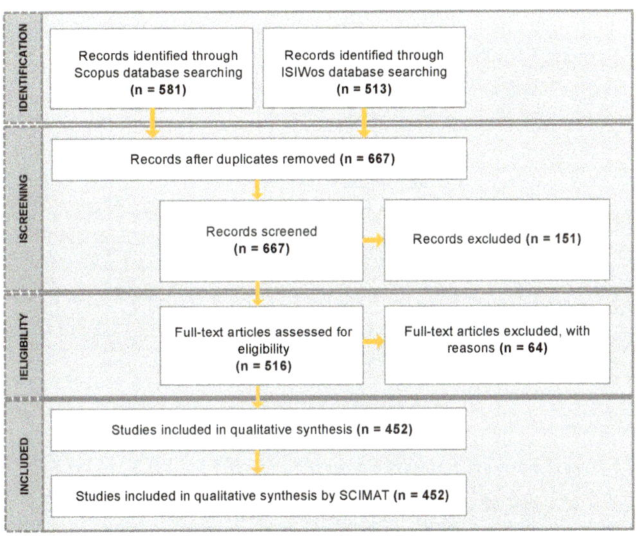

Figure 3. PRISMA flow diagram.

Figure 4. Strategic diagrams by period.

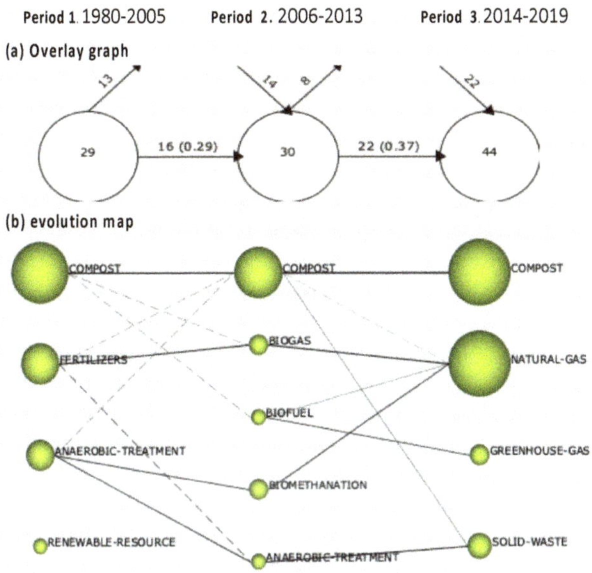

Figure 5. Overlay graph (**a**) and evolution map (**b**) of the research field by periods.

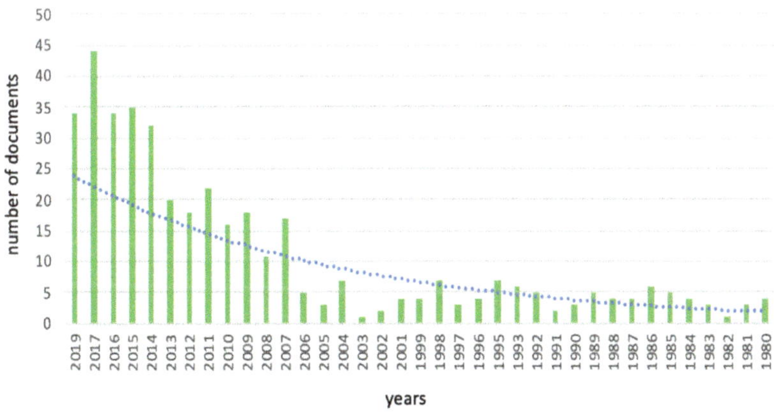

Figure 6. Documents per year.

Table 1. Documents by search strings.

Search String *	Records Scopus	Records ISIWoS
"Organic waste" AND "municipal solid waste" AND "biogas"	250	235
"Organic waste" AND "municipal solid waste" AND "biomethane"	19	21
"Organic waste" AND "municipal solid waste" AND "natural gas"	18	15
"Organic waste" AND "municipal solid waste" AND "network"	20	9
"Organic waste" AND "municipal solid waste" AND "compost"	274	233
Total Records	581	513

* Search completed 31 December 2019.

Table 2. Performance analysis by period.

Name	No. of Documents	No. of Citations	h-Index	Centrality	Density
Period 1 (1980–2005)					
Compost	10	116	5	7.94	32.98
Fertilizers	6	26	2	12.64	10.52
Anaerobic treatment	4	25	2	10.69	15
Renewable resource	1	0	0	3.21	25
Period 2 (2006–2013)					
Compost	8	36	2	8.75	9.63
Biogas	2	26	2	6.25	40
Biofuel	1	2	1	0	150
Bio methanation	2	1	1	0	50
Anaerobic treatment	1	2	1	0	12.5
Period 3 (2014–2019)					
Compost	11	31	4	19.47	14.05
Natural gas	11	32	3	15.54	39.31
Greenhouse gases	2	6	1	8.64	39.58
Solid waste	3	2	1	6.51	4.86

Table 3. Main publications contributing to the research field.

	Name	Total Number of Citations of the Journal on This Study	No. of Documents
1	Compost Science and Utilization	162	12
2	Agricultural Wastes	132	10
3	BioCycle	93	38
4	Environmental Engineering and Management Journal	66	10
5	Journal of Material Cycles and Waste Management	45	10
6	Polish Journal of Environmental Studies	21	6
7	Acta Horticulturae	17	11
8	Agronomy Research	16	7
9	Desalination and Water Treatment	11	5
10	Journal of Ecological Engineering	3	6

Table 4. Authors with more than ten published studies in the research field.

	Name	No. of Documents	Total Citations in This Work	h-Index
1	Dubrovskis, V.	4	3	3
2	Li, Y.	4	12	2
3	Kumar, A.	4	8	5
4	Cioabla, A.E.	3	2	4
5	Li, J.	3	12	9
6	Liu, H.	3	6	26
7	Kacprzak, M.	3	33	15
8	Zhao, Y.	3	7	15
9	Xi, B.	3	7	31
10	Guerrini, O.	3	5	4

3.1. *Systematic Literature Review*

The results obtained by applying the methodology of the SLR are presented below, including the definition of the research questions, the search process, the PRISMA flow diagram (Figure 3) and the data and search results collection.

3.1.1. Planning and Formulation of the Problem

The research questions were determined before starting the search. The SLR of this study addressed the following research questions (RQ):

- RQ1: What is the objective of this review?
- RQ2: What is the status of this study field?
- RQ3: Towards what topics is the field of research evolving?
- RQ4. What research topics are being addressed?
- RQ5. Who is leading the research?
- RQ6. What are the limitations of the current research?
- RQ7. Where are these papers published (e.g., journals, conferences)?

3.1.2. Selection of the Database, Keywords and Search Strings

Firstly, the Scopus and ISIWoS databases were selected because of their large numbers of international scientific publications and high impact techniques for any discipline. Secondly, the keywords related to OFMSW recycling were identified. Next, an advanced search was carried out in the field "Title/Abstract/Keyword" with the identified keywords and using the five search chains defined in Table 1. Finally, a total of 581 bibliographic records were identified for Scopus and 513 bibliographic records, for ISIWoS.

3.1.3. Selection of the Literature

Once these documents had been compiled, the guidelines of the PRISMA flow chart were applied to show the number of relevant documents (Figure 3). It can be seen that a total of 1094 bibliographic records were retrieved from the two selected databases. After removing 427 duplicates, 177 of the remaining 667 records were excluded based on their titles and abstracts; notes and errata were also excluded. The remaining 490 records were examined at the full-text level, which led to the exclusion of 86 additional records that did not cover the topics included in this study.

3.1.4. Identification of the Time Horizon

The recovery time interval of the literature was established from 1980 to 2019. To analyse trends in publication patterns, this time interval of the study was divided into three periods, considering both several relevant milestones and the number of documents selected. As a result, the following three periods were established:

- First period (1980–2005). Article 5 of Directive 31/99 (EUC, 1999) established that Member States had to develop a national strategy to reduce biodegradable waste disposal to landfill no later than two years after its publication. This strategy had to include measures to achieve the objectives established through recycling, composting, biogasification or the valorisation of materials/energy. Thus, not later than 2006, biodegradable municipal waste disposal in landfills had to be reduced to 75% of the total amount (by weight) of municipal biodegradable waste generated in 1995, and by 2016, to 35%. For this reason, the beginning of the first period established corresponds to the date of the first document selected; the end, the year 2005, corresponds to the first date established to reduce the organic matter deposited in landfill.
- Second period (2006–2013). The end of this second stage corresponds to the date of the second commitment period established in the Kyoto Protocol (2013), which ends in the year 2020. The Doha Amendment also applies, according to which the participating countries committed to reducing emissions by at least 18%, compared to 1990 levels. In the case of the EU, it undertook to reduce emissions by 20% below 1990 levels (United Nations, 1998).
- Third period (2014–2019). The end of this period corresponds to the date of the last articles included in the study, that is, the past year.

3.2. *Bibliometric Analysis: Science Mapping and Analysis Performance*

3.2.1. Science Mapping and Strategic Diagrams

From the analysis of the evolution of the strategic diagrams, the change in the development of the recycling of OFMSW can be seen, as well as the main milestones and inflection points. For the three periods considered (1980–2005, 2006–2013 and 2014–2019), and in order to analyse the temporal evolution, Figure 4 represents the strategic diagrams, showing the sizes of the spheres proportional to the numbers of documents published associated with each research topic. In addition, Table 2 shows the measures of performance obtained for each topic and period in terms of the number of documents, h-index, values of centrality and density. An analysis of these results, for each period, is discussed below.

- *First period (1980–2005).* According to the strategic diagram of Figure 4a, the following four main research topics can be found in the 97 papers published in this period: compost, fertilisers, anaerobic treatment and renewable resources. Two of them are considered motor themes (*compost* and *anaerobic treatment*); one, transversal (*renewable resource*); and, finally, another, a basic one (*fertilizers*). The performance analysis for each topic (Table 2) complements the information provided by the strategic diagram; it may be observed that *compost* and *fertilizer* are the themes with a significant impact rate; they receive more than 100 citations and obtain higher h-indices compared to the remaining themes. These research topics show that the first treatments of the organic fraction of urban waste were aimed at producing compost to be used as fertiliser, applying simple technologies for the aerobic stabilisation of the biodegradable fraction.
- *Second period (2006–2013).* According to the strategic diagram of Figure 4b, in the 127 papers published in this period, an increase in research topics is seen, rising to five: *compost, biogas, biofuel, biomethanization* and *anaerobic treatment*. One of them is considered a motor theme (*biogas*); two, transversal (*biofuel* and *biomethanization*); and, finally, two of them are basic themes (*anaerobic treatment* and *compost*). The performance measures included in Table 2 reveal that *compost* and *biogas* are the most noted research topics. They obtained an important impact rate and achieved higher h-indices in comparison with the remaining topics. It can be seen that the application of the waste management hierarchy and the obligation to achieve the established objectives to reduce the percentage of biodegradable waste disposal in landfill have led to an evolution of the treatments towards anaerobic stabilisation or biomethanisation technologies in order to obtain a biogas that can be used to produce energy, as well as a digestate that, after a composting phase, can be used in agriculture.
- *Third period (2014–2019).* Finally, in the last period (Figure 4c), there is a greater number of documents (245), so it is possible to differentiate four research topics: *compost, natural gas, greenhouse gases* and *solid waste*. Three of these research topics are considered motor themes (*compost, natural gas* and *greenhouse gases*), and only one of them is classified as declined (*solid waste*). The performance measures highlight two research topics: compost and natural gas; they show an important impact rate and achieve higher h-indices compared to the remaining topics. Although the compost theme appears as the main one, research focused on the use of biomethane in natural gas networks emerges as a way to reduce Greenhouse Gas (GHG). This is a clear example of the important role of waste management in the decarbonisation of the energy system.

3.2.2. Science Mapping, Overlay Graph and Thematic Evolution Map

The systematic analysis of the literature has shown the change in the development of the treatments for recycling OFMSW, as well as the main milestones and inflection points. Next, it was considered interesting to carry out a joint analysis of the evolution of the keywords and the thematic evolution of the field of the investigation. The results are shown in Figure 5 and discussed below.

The number of keywords per period and their evolution have been represented in Figure 5a, as well as the number of incoming and outgoing keywords, and the number and percentage of keywords that remain from one period to the next. It can be seen that the number of keywords increases over the periods, in parallel with the rise in the number of documents over the years. Thus, the number of keywords increased from 29 to 44 between the first and last periods, which meant a growth of 51.7%; this result indicates that the field of research is diversifying and continuing to increase, meaning that it is not yet a consolidated field. The increasing number of words and keywords shared between successive periods proves the growing thematic diversity of the field of research on the recycling of OFMSW.

Figure 5b shows the thematic evolution of the field of research thanks to the analysis of its origins and its interrelations. The thickness of the lines represents the strength of the association measured by the inclusion index. The analysis of the graph from the point of view of the number of documents shows that the four thematic groups of the first period (1980–2005) have progressed towards different concepts; thus, the compost theme appeared with the largest number of central documents in 1980–2005, evolving into the topics of biogas, biofuel and new compost in 2006–2013; in the last period (2014–2019), compost appears again with the largest number and has evolved into the topics of natural gas and solid waste. On the other hand, the fertilisers thematic group has evolved towards anaerobic treatment, compost and biogas in the period of 2006–2013, to finally evolve to natural gas in the last period; this result reveals again the evolution of the by-products of the treatments used in the stabilisation of organic matter.

Finally, it is noteworthy that the compost thematic group of the first period was kept with the same label in the second and third periods but with a greater number of central documents published during the last one (2006–2019). In the same way, the anaerobic treatment thematic group of the first period was maintained with the same label in the following period, although with a lower number of documents, evolving towards the solid waste thematic group during the last period.

3.2.3. Performance Analysis

After the SLR was performed, a total of 452 documents published within the time horizon (1980–2019) were obtained. Finally, the following configuration in SciMAT for the bibliometric analysis was established: (i) the word as the unit of analysis, (ii) the analysis of co-occurrence to build the networks, (iii) the index of equivalence to measure the similarity to standardize the networks, and (iv) the k-means clustering algorithm to detect the themes; finally, documents were analysed taking into account the year of publication, journals cited, authors and number of citations. The results obtained are summarised below:

- Documents per year. Figure 6 shows the distribution by year of the 452 publications selected. It is worth stressing that, in general terms, the number of studies was not high, except in the year 2017, when it exceeded 40. Before 2007, no more than eight publications related to the research field analysed were observed per year; however, since 2007, a continuing increase in the number of articles can be seen. This result highlights the obligation to apply Article 4 of Directive 2006/12/EC of the European Parliament and of the Council of 5 April 2006 on waste, which established the obligation to apply the waste management hierarchy; it indicates an order of preference for management hierarchy that reduce the production of waste and capture the progression of a material or product through consecutive phases of waste management, and it includes, in this order: prevention; preparation for reuse; recycling; another type of valorisation, for example, energy recovery; and elimination.
- Most relevant journals. A total of 192 journals were identified in this analysis. Table 3 shows the journals of 28.5% of the documents analysed, classified in descending order according to the number of citations of the document. Most of these research journals focus on the application of processed organic solid waste, biogas utilisation, energy policy and soil science, among others. Table 3 also includes the most cited publications in each journal; it should be noted that the numbers of publications and citations are closely related, except for the BioCycle journal; that is, the most prolific sources are those with the greatest impact in the research field.
- Documents by author. A systematic literature review and performance analysis allowed the identification of a total of 1329 authors who have published on the theme addressed by the study's objective. Table 4 shows those authors with more than three published studies, as well as the total number of documents published and citations received; the h-index (Hirsch index), has been also included as a measure of the authors' professional quality, taking into account the number of times that their scientific articles have been cited [39]. According to the information analysis, Dubrovskis, V., Li, Y. and Kumar, A. have published the most articles on them, although Xi, B. has the highest h-index.

4. Conclusions

This document shows an analysis of the scientific literature that addresses the treatment of the biodegradable fraction of urban waste from 1980 to 2019. To do this, a transparent, rigorous and reproducible research procedure was applied to a collection of 452 articles published in indexed journals in the ISI Web of Science (ISIWoS) and Scopus database, with peer review before publication. Tendencies were analysed, considering an overview and a more specific analysis of three different time intervals during the period under review (1980–2005, 2006–2013 and 2014–2019).

The study has shown that the application of technologies that allow the use of the OFMSW as a biological nutrient in the framework of the circular economy is relevant, especially in recent years. In fact, the results show a regular rise in the number of studies published since 1980; this rise is more significant since 2014, when policies aimed at reducing emissions to the atmosphere through the development of renewable energies were increased. Although the treatments of this fraction have been applied for years, the scientific field is still evolving; this may be explained by the evolution of the legal framework, which imposes the need to continue working on the implementation of strategies to develop a model of a circular economy. One of the basic pillars of this new framework is the use of bioproducts, such as OFMSW, as a renewable energy resource.

The systematic analysis of the literature has shown the evolution of recycling treatment for OFMSW, as well as the main landmarks and inflection points. On the other hand, the strategic diagrams show the interest of researchers in different key issues, which have progressed from simple treatments, just applied to produce compost from the biodegradable fraction of municipal waste, up to its use to produce biofuel; this last application is within the framework of policies aimed at promoting the use of renewable energies that reduce the emissions of greenhouse gases.

This document has shown that emerging studies in this area are being conducted on the production of biomethane, called green gas, for application as fuel in vehicles or for injection into natural gas networks. Although these applications are currently in use, they are still not competitive, but their contribution to their circular economy as a sustainable alternative to waste disposal, as well as to reduce greenhouse gas emissions, is already clear. Biomethane is thus positioned as a fundamental element in the energy transition that will contribute to a decarbonised energy system in order to comply with global objectives.

The above findings provide researchers and professionals in the OFMSW recycling field with a detailed understanding of the status quo, predicting its dynamic directions; in consequence, this study is a valuable contribution to research concerning this field.

Author Contributions: Conceptualisation, J.M.F.-G. and J.M.-P.; methodology, C.D.-L.; software, C.D.-L.; analysis of results, M.Z.; writing—original draft preparation, C.D.-L. and M.Z.; writing—review and editing, J.M.-P. All authors have read and agreed to the published version of the manuscript.

Funding: This research received no external funding.

Acknowledgments: The authors appreciate the support of the research group TEP-968 (Technologies for Circular Economy) of the University of Granada.

Conflicts of Interest: The authors declare no conflict of interest.

References

1. Ragazzi, M.; Rada, E.C. Effects of recent strategies of selective collection on the design of municipal solid waste treatment plants in Italy. *WIT Trans. Ecol. Environ.* **2008**, *109*, 613–620. [CrossRef]
2. Abdel-Shafy, H.I.; Mansour, M.S.M. Solid waste issue: Sources, composition, disposal, recycling, and valorization. *Egypt. J. Pet.* **2018**, *27*, 1275–1290. [CrossRef]
3. Edjabou, M.E.; Jensen, M.B.; Götze, R.; Pivnenko, K.; Petersen, C.; Scheutz, C.; Astrup, T.F. Municipal solid waste composition: Sampling methodology, statistical analyses, and case study evaluation. *Waste Manag.* **2015**, *36*, 12–23. [CrossRef] [PubMed]

4. Johnstone, N.; Labonne, J. Generation of Household Solid Waste in OECD Countries: An Empirical Analysis Using Macroeconomic Data. *Land Econ.* **2004**, *80*, 529. [CrossRef]
5. Al-Khatib, I.A.; Monou, M.; Abu Zahra, A.S.F.; Shaheen, H.Q.; Kassinos, D. Solid waste characterization, quantification and management practices in developing countries. A case study: Nablus district – Palestine. *J. Environ. Manag.* **2010**, *91*, 1131–1138. [CrossRef] [PubMed]
6. Hoornweg, D.; Bhada-Tata, P.; Kennedy, C. Waste production must peak this century. *Nature* **2013**, *502*, 615–617. [CrossRef] [PubMed]
7. Lim, S.L.; Lee, L.H.; Wu, T.Y. Sustainability of using composting and vermicomposting technologies for organic solid waste biotransformation: Recent overview, greenhouse gases emissions and economic analysis. *J. Clean. Prod.* **2016**, *111*, 262–278. [CrossRef]
8. Andreottola, G.; Ragazzi, M.; Foladori, P.; Villa, R.; Langone, M.; Rada, E.C. The unit intregrated approch for OFMSW treatment. *UPB Sci. Bull. Ser. C Electr. Eng.* **2012**, *74*, 19–26.
9. Sheets, J.P.; Yang, L.; Ge, X.; Wang, Z.; Li, Y. Beyond land application: Emerging technologies for the treatment and reuse of anaerobically digested agricultural and food waste. *Waste Manag.* **2015**, *44*, 94–115. [CrossRef]
10. Tuomela, M.; Vikman, M.; Hatakka, A.; Itävaara, M. Biodegradation of lignin in a compost environment: A review. *Bioresour. Technol.* **2000**, *72*, 169–183. [CrossRef]
11. Farrell, M.; Jones, D.L. Critical evaluation of municipal solid waste composting and potential compost markets. *Bioresour. Technol.* **2009**, *100*, 4301–4310. [CrossRef]
12. Yang, L.; Ge, X.; Wan, C.; Yu, F.; Li, Y. Progress and perspectives in converting biogas to transportation fuels. *Renew. Sustain. Energy Rev.* **2014**, *40*, 1133–1152. [CrossRef]
13. Gabhane, J.; William, S.P.; Bidyadhar, R.; Bhilawe, P.; Anand, D.; Vaidya, A.N.; Wate, S.R. Additives aided composting of green waste: Effects on organic matter degradation, compost maturity, and quality of the finished compost. *Bioresour. Technol.* **2012**, *114*, 382–388. [CrossRef] [PubMed]
14. Nakhshiniev, B.; Gonzales, H.B.; Yoshikawa, K. Hydrothermal Treatment of Date Palm Lignocellulose Residue for Organic Fertilizer Conversion: Effect on Cell Wall and Aerobic Degradation Rate. *Compost Sci. Util.* **2012**, *20*, 245–253. [CrossRef]
15. Wu, Y.-R.; He, J. Characterization of anaerobic consortia coupled lignin depolymerization with biomethane generation. *Bioresour. Technol.* **2013**, *139*, 5–12. [CrossRef] [PubMed]
16. Jiménez, E.I.; Garcia, V.P. Composting of domestic refuse and sewage sludge. II. Evolution of carbon and some "humification" indexes. *Resour. Conserv. Recycl.* **1992**, *6*, 243–257. [CrossRef]
17. Cofie, O.; Kone, D.; Rothenberger, S.; Moser, D.; Zubruegg, C. Co-composting of faecal sludge and organic solid waste for agriculture: Process dynamics. *Water Res.* **2009**, *43*, 4665–4675. [CrossRef]
18. Bernal, M.P.; Alburquerque, J.A.; Moral, R. Composting of animal manures and chemical criteria for compost maturity assessment. A review. *Bioresour. Technol.* **2009**, *100*, 5444–5453. [CrossRef]
19. Bustamante, M.A.; Suárez-Estrella, F.; Torrecillas, C.; Paredes, C.; Moral, R.; Moreno, J. Use of chemometrics in the chemical and microbiological characterization of composts from agroindustrial wastes. *Bioresour. Technol.* **2010**, *101*, 4068–4074. [CrossRef]
20. Bustamante, M.A.; Alburquerque, J.A.; Restrepo, A.P.; de la Fuente, C.; Paredes, C.; Moral, R.; Bernal, M.P. Co-composting of the solid fraction of anaerobic digestates, to obtain added-value materials for use in agriculture. *Biomass Bioenergy* **2012**, *43*, 26–35. [CrossRef]
21. Basso, D.; Weiss-Hortala, E.; Patuzzi, F.; Castello, D.; Baratieri, M.; Fiori, L. Hydrothermal carbonization of off-specification compost: A byproduct of the organic municipal solid waste treatment. *Bioresour. Technol.* **2015**, *182*, 217–224. [CrossRef] [PubMed]
22. Circular Economy Strategy - Environment - European Commission. Available online: http://ec.europa.eu/environment/circular-economy/index_en.htm (accessed on 23 April 2019).
23. Kothari, R.; Tyagi, V.V.; Pathak, A. Waste-to-energy: A way from renewable energy sources to sustainable development. *Renew. Sustain. Energy Rev.* **2010**, *14*, 3164–3170. [CrossRef]
24. Noyons, E.C.M.; Moed, H.F.; Luwel, M. Combining mapping and citation analysis for evaluative bibliometric purposes: A bibliometric study. *J. Am. Soc. Inf. Sci.* **1999**, *50*, 115–131. [CrossRef]
25. van Nunen, K.; Li, J.; Reniers, G.; Ponnet, K. Bibliometric analysis of safety culture research. *Saf. Sci.* **2018**, *108*, 248–258. [CrossRef]
26. Roig-Tierno, N.; Gonzalez-Cruz, T.F.; Llopis-Martinez, J. An overview of qualitative comparative analysis: A bibliometric analysis. *Suma Negocios* **2017**, *2*, 15–23. [CrossRef]

27. Small, H. Visualizing science by citation mapping. *J. Am. Soc. Inf. Sci.* **1999**, *50*, 799–813. [CrossRef]
28. Albort-Morant, G.; Ribeiro-Soriano, D. A bibliometric analysis of international impact of business incubators. *J. Bus. Res.* **2016**, *69*, 1775–1779. [CrossRef]
29. Henderson, M.; Shurville, S.; Fernstrom, K. The quantitative crunch. *Campus-Wide Inf. Syst.* **2009**, *26*, 149–167. [CrossRef]
30. Liberati, A.; Altman, D.G.; Tetzlaff, J.; Mulrow, C.; Gøtzsche, P.C.; Ioannidis, J.P.A.; Clarke, M.; Devereaux, P.J.; Kleijnen, J.; Moher, D. The PRISMA Statement for Reporting Systematic Reviews and Meta-Analyses of Studies That Evaluate Health Care Interventions: Explanation and Elaboration. *PLoS Med.* **2009**, *6*, e1000100. [CrossRef]
31. Martínez-Aires, M.D.; López-Alonso, M.; Martínez-Rojas, M. Building information modeling and safety management: A systematic review. *Saf. Sci.* **2018**, *101*, 11–18. [CrossRef]
32. Antman, E.M.; Lau, J.; Kupelnick, B.; Mosteller, F.; Chalmers, T.C. A Comparison of Results of Meta-analyses of Randomized Control Trials and Recommendations of Clinical Experts. *JAMA* **1992**, *268*, 240. [CrossRef] [PubMed]
33. da Silva Serapião Leal, G.; Guédria, W.; Panetto, H. Interoperability assessment: A systematic literature review. *Comput. Ind.* **2019**, *106*, 111–132. [CrossRef]
34. Kitchenham, B.; Charters, S. Procedures for Performing Systematic Literature Reviews in Software Engineering. *Keele Univ. Durham Univ. UK* **2007**.
35. Liberati, A.; Altman, D.G.; Tetzlaff, J.; Mulrow, C.; Gøtzsche, P.C.; Ioannidis, J.P.A.; Clarke, M.; Devereaux, P.J.; Kleijnen, J.; Moher, D. The PRISMA statement for reporting systematic reviews and meta-analyses of studies that evaluate healthcare interventions: Explanation and elaboration. *BMJ* **2009**, *339*, b2700. [CrossRef]
36. Coulter, N.; Monarch, I.; Konda, S. Software engineering as seen through its research literature: A study in co-word analysis. *J. Am. Soc. Inf. Sci.* **1998**, *49*, 1206–1223. [CrossRef]
37. Alcaide–Muñoz, L.; Rodríguez–Bolívar, M.P.; Cobo, M.J.; Herrera–Viedma, E. Analysing the scientific evolution of e-Government using a science mapping approach. *Gov. Inf. Q.* **2017**, *34*, 545–555. [CrossRef]
38. Hirsch, J.E. An index to quantify an individual's scientific research output. *Proc. Natl. Acad. Sci. USA* **2005**, *102*, 16569–16572. [CrossRef]
39. Schreiber, M. Restricting the h-index to a publication and citation time window: A case study of a timed Hirsch index. *J. Informetr.* **2015**, *9*, 150–155. [CrossRef]

sustainability

MDPI

Article

Determination of the Optimal Operative Conditions for the Torrefaction of Olive Waste Biomass

Jaime Martín-Pascual [1], Joaquín Jódar [2], Miguel L. Rodríguez [3] and Montserrat Zamorano [1,*]

[1] Department of Civil Engineering, Campus Fuentenueva, University of Granada, 18071 Granada, Spain; jmpascual@ugr.es
[2] Department of Mathematics, Campus Las Lagunillas, University of Jaén, 23071 Jaén, Spain; jjodar@ujaen.es
[3] Department of Applied Mathematics, Campus Fuentenueva, University of Granada, 18071 Granada, Spain; miguelrg@ugr.es
* Correspondence: zamorano@ugr.es

Received: 9 June 2020; Accepted: 6 August 2020; Published: 10 August 2020

Abstract: The need for new energy sources and the problems associated with waste in the agroforestry industry are an opportunity for the recovery of this waste. For the use of this agricultural waste as energy, different pretreatments, such as torrefaction, can be carried out. Torrefaction is a thermochemical treatment involving energetic densification of biomass at temperatures ranging from 200 to 300 °C under an inert and anaerobic environment. This study developed a numerical model to evaluate the effect of temperature and residence time of torrefaction on biomass from olive tree waste to determine optimum operative conditions for the process. Four temperatures and four residence times, in the operation range of the process, were tested to determine the weight loss and the higher heating values (HHVs) of the torrefied sample. From these data, a numerical model was developed to infer the complete behavior of the process in the temperature range between 200 and 300 °C and in the residence time range of a few minutes to 2 h. The HHV of the torrefied sample increased at a temperature between 200 and 275 °C. However, from 275 to 300 °C, there was an HHV decrease. The effect of the residence time depended on the torrefaction temperature. At low temperatures, there were no statistically significant differences, although an increase of HHV was detected under 120 min. However, at 250 °C this effect was reversed, and statistically significant differences were not observed between 30 and 120 min. Overall, the increase of temperature in the torrefaction process reduces the residence time needed to achieve the maximum HHV. As a result, the optimum conditions of torrefaction for this biomass were, approximately, 275 °C and 30 min of residence time. This reaction yielded an optimum 5830 cal/g HHV.

Keywords: biomass; olive waste; energetic densification; pretreatment; torrefaction

1. Introduction

Environmental issues and climate change will require our society to transition to more sustainable means. To achieve this transition, the development of a new concept of waste disposal and energy creation must be considered. Efficient valorization of agricultural waste is a key factor in developing new strategies for the circular economy [1], and renewable energy sources can be used worldwide to mitigate the impact of global warming and to decrease the high dependence on fossil fuels in the energy market [2,3]. Renewable energy has environmental benefits, since bioenergy can not only reduce carbon dioxide emissions but also decrease the environmental impact caused by organic wastes and the economic development of rural areas. Biomass is one emerging and fundamentally important source of renewable energy [4]. Biomass-derived energy is often preferred over other renewable alternatives, including wind and solar power, due to its higher and decentralized availability [5].

Biomass is a primary source of renewable carbon that can be used as feedstock for biofuel production. Using biomass as an energy source allows for energy independence [6] because it can be converted into fuels and chemicals through thermochemical and biochemical processes, making it a potential alternative to fossil fuels [7,8]. There is considerable potential for bioenergy from several sources, since a wide range of feedstocks can be used for bioenergy generation. These sources include energy crops, biomass residues, and organic wastes [4]. Lignocellulosic biomass is a renewable energy source with a carbon-neutral cycle and relatively low cost of production, originating from energy plantations or residues from primary or industrial processing of crops and forest products [9].

While in northern Europe, it is common to use wood biomass, such as bark, wood chips, and sawdust, the Mediterranean area has great potential from agricultural residues, namely the olive oil sector [4]. Olives are the most extensively cultivated fruit crop in the world and are particularly widespread throughout the Mediterranean region. Olives play an important role in the rural economy, local heritage, and environment protection of the Mediterranean region [10]. Furthermore, in 2014, the five largest olive oil producer countries were Spain, Italy, Greece, Tunisia, and Morocco [1]. There is consequently a considerable amount of waste generated from the olive industry in these countries. In this context, several options of valorization may be of interest, especially given the amounts produced and environmental impacts caused [11].

The residual biomass produced in the olive sector is the result of the large quantity of olive groves and olive oil manufacturers that generate byproducts with a potentially high energy content [12]. Moreover, olive tree waste production takes place for a short period of the year, so large amounts of waste accumulate in a short period of time. All these facts hamper the use of raw olive biomass due to the difficulty of storing, transporting, and grinding it [13]. One of the main alternatives to enhance the energy quality of the biomass and to support its increased use as a fuel source is the application of post-harvest treatments [9]. Many methods used to extract energy from lignocelluloses have been developed, including combustion, pyrolysis, gasification, and torrefaction [14]. Torrefaction can help to overcome some of the above-mentioned limitations by converting biomass into an upgraded solid material with increased energy density and decreased oxygen content, which is, therefore, more suitable for energy generation [4]. Elemental analysis done by Martin-Lara et al. revealed that the composition of olive tree pruning moved from lignocellulosic biomass to coal during the torrefaction [15].

Torrefaction is a thermochemical technology used to treat biomass at temperatures ranging from 200 to 300 °C under an inert, anaerobic environment, such as nitrogen or argon [13,16,17]. Although there is a small loss of carbon from biomass during torrefaction, a large quantity of oxygenated compounds is lost as well [14], thus enabling energy densification of biomass and biomass homogenization [18,19].

Torrefaction is a mild pyrolysis process that may overcome some of the previously mentioned limitations, thus improving the quality of biomass feedstock [20,21]. During torrefaction, biomass is converted into an upgraded solid material with increased energy density and decreased oxygen content, which is more suitable for energy generation [4]. Torrefaction also contributes towards addressing the challenges associated with the supply chain management (regarding storage, handling, and transportation costs) [3].

Torrefied biomass has a higher heating value than raw biomass, mainly due to the reduction of moisture content and the atomic O/C and H/C ratios [3]. Therefore, torrefaction has been recommended as an efficient way to enhance solid biofuel properties through water removal, reduction of the hygroscopic range, increased grindability [6] and resistance to degradation, among other properties [22]. Torrefied material is comparable with a low-rank coal. It still retains some characteristic properties from its original biomass yet has a higher energy content and better stability against microbial degradation due to improved hydrophobic properties [4].

During torrefaction, cell walls are degraded, producing fuel in a solid form [6] with intermediate characteristics between raw biomass and charcoal. These characteristics include dark color, high carbon and energy contents, and low equilibrium moisture content [22–24]. During the process, other volatile products (carbon dioxide, carbon monoxide, and possible traces of acetic acid, hydrogen, and methane) and condensable and non-condensable gases (water vapor, acetic acid, furfural, formic acid, methanol, lactic acid, phenol) are also produced [4]. The volatiles eliminated during torrefaction come from partial and selective degradation of the lowest calorific fraction of the biomass composition, such as hemicelluloses and some extractives [9,25].

Torrefaction is influenced by many parameters, including biomass composition, physical properties, and operating conditions [6,19]. Temperature and residence time are the two most important parameters that influence the torrefaction process [26]. With the increase of temperature or residence time, mass efficiency decreases in the different chemical reactions that occur during the process itself [27]. Considering these parameters, the aim of this study was to evaluate the effect of temperature (200, 250, 275, and 300 °C) and residence time (0, 30, 60, and 120 min) of torrefaction on biomass from olive tree waste in relation to its weight loss and higher heating value (HHV). With these results, this study determined the optimal conditions for increasing the HHV of olive waste as a renewable energy source.

2. Materials and Methods

2.1. *Characteristics of the Biomass*

The biomass from olive trees used in this research included leaves and small tree branches originating from agricultural activities in Granada (Spain).

2.2. *Torrefaction Process*

Biomass was dried at 105 °C for 24 h prior to torrefaction to remove any residual water in the biomass (Figure 1a). It was needed to reduce the granulometric values of the raw materials, as a particle size smaller than 0.5 mm was necessary for thermogravimetric analysis and for ensuring the heat transfer rate.

The torrefaction process of the biomass in this study was performed with a Mettler Toledo TGA/DSC 1 thermogravimetric analyzer (Figure 1b) under inert atmosphere using 900 μL alumina crucibles as previously done by Harun et al. [28]. The initial mass of the samples was kept between 7.5 and 8 mg to avoid any possible effect on mass and heat transfer during the process. First, samples were heated from ambient temperature to the torrefaction temperature (200, 250, 275, or 300 °C) at heating rates of 10 °C/min and with 10 mL/min of nitrogen gas, as in Nakason et al. [29]. To analyze and model the effect of residence time in the torrefaction process, four conditions of this variable were tested. In the first condition, once the temperature was reached, the process was stopped (0 min of residence time), as in the study by Arias et al. [19]. In the other three conditions, when the target temperature was achieved, this was maintained during the tested residence time (for 30, 60, and 120 min). During the process, the weight loss of torrefied biomass can be determined from the percentage of the initial mass that remains at the end of the torrefaction procedure (mass yield). At the end, the HHV of torrefied sample (Figure 1c) was measured.

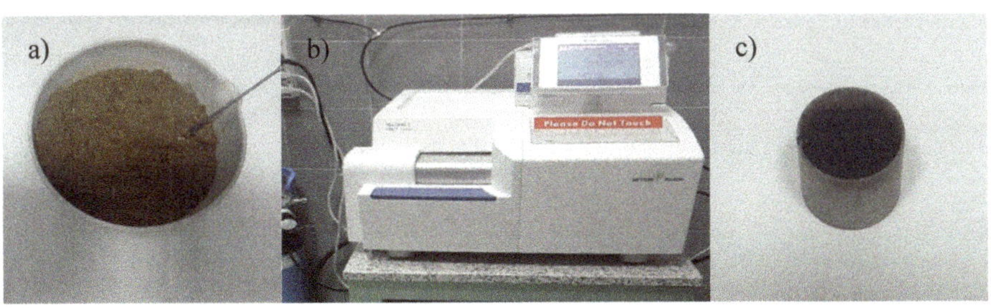

Figure 1. Torrefaction process: (**a**) Raw material, (**b**) thermogravimetric analyzer used for torrefaction process, and (**c**) torrefied sample.

2.3. Moisture and Ash Content

Three samples were milled from raw material to determine their moisture content in a drying oven (105 ± 2 °C) to a constant weight and according to CEN/TS 14774-2:2004. The ash content was determined using the TGA/DSC 1 thermogravimetric analyzer (METTLER TOLEDO, Columbus, OH, United States). After the torrefaction procedure, the loss of ignition (at 550 °C) of the torrefied biomass samples was measured according to CEN/TS 14775:2004.

2.4. Determination of Higher Heating Value (HHV)

Three samples from raw material and their corresponding torrefied biomass samples were tested under the different conditions and milled to determine their heating values with a bomb calorimeter IKA C 2000, according to UNE 164001:2005 EX.

2.5. Statistical Analysis

Data obtained throughout this study were analyzed using SPSS 20 for Windows. A least significant difference (LSD) test was used to measure the differences between the weight loss and HHV obtained under the different operational conditions of temperature and residence time. An analysis of variance (ANOVA) was used to assess the homogeneity of variance, with a significance level of 5% ($p < 0.05$).

2.6. Numerical Model

The data experimentally obtained were used for determining two approximating functions, one for mass yield and the other one for HHV.

For a given set of data (measurements and locations at which these measurements were obtained), the approximation procedure usually tries to determine a function ("approximating function") that is a good fit for the given data. It is considered that this good fit is achieved if the values provided by the approximating function exactly match the given measurements at the corresponding locations (or at least are close to these measurements). Once the approximating function is determined, information can also be deduced about the studied problem at locations different from those at which the measurements were obtained.

There are several techniques to determine this approximating function. The method of least squares is useful for obtaining an approximation of a set of points by analytic expressions. It is also a recommended method for approximating the problem when the available number of points is small, as was the case in this study. Its name, least squares, is due to the fact that this mathematical procedure finds the best-fitting function to a given set of points by minimizing some errors, typically the sum-of-the-squares of the residual errors. In other words, this least square deviation is reached by the function for which that minimum is achieved [30].

Additionally, approximation by radial basis functions has proved to be very useful in numerical analysis, in numerical treatment of differential, integral, and partial differential equations, in statistics, and has found applications in science, engineering, economics, biology, medicine, etc. To approximate mass yield and HHV from the data experimentally obtained, an approximant expressed as a finite linear combination of a certain radial basis function and its translations was sought. In order to do this approximation, the multiquadric function given by the expression $\phi(r) = \sqrt{1+(\varepsilon r)^2}$, $r \geq 0$, was chosen as the basis function, but there are other possibilities. A wide range of radial basis functions can be found in the literature [31,32]. The parameter $\varepsilon \geq 0$ that appears in the above expression is a shape parameter.

More precisely, the formulation of the problem was the following:

Given n points $(x_i, y_i, z_i) \in R^3$, $i = 1, \ldots, n$, a function $s(x, y)$ that approximated the given scalar values z_i at the points (x_i, y_i) in the least-squares sense was sought. Specifically, expression (1), where f_i, $i = 1, \ldots, n$, constituted a set of linear independent functions on R^2 and, a_i, $i = 1, \ldots, n$, constituted a set of real coefficients to be determined, was set.

$$s(x,y) = \sum_{i=1}^{n} a_i f_i(x,y) \quad (1)$$

To this end, the error functional according to the expression (2) was defined, and the following minimization problem was posed: min $E(a_1, \ldots, a_n)$.

$$E(a_1, \ldots, a_n) = \sum_{j=1}^{n} \left(s\left(x_j, y_j\right) - z_j\right)^2 \quad (2)$$

For all our examples, $s(x, y)$ was assumed to have the form of expression (3), where $\|.\|$ is the Euclidean norm on R^2 and $\phi : [0, \infty) \to R$, is the basis function. In our particular case, $k = n$ was chosen. As mentioned, the function $s(x, y)$ is known when the a_i values are determined, and this is basically done by solving a linear equation system. Namely, the critical points of the error functional, that is, those points $(a_1, \ldots, a_n)$ for which all the first-order partial derivatives of $E(a_1, \ldots, a_n)$ are zero, were firstly computed. Then the second derivative test was used to check if the obtained critical points were indeed minimizers of the functional E. This method is described in detail in many books [30].

$$s(x,y) = \sum_{i=1}^{k} a_i \phi(\| (x,y) - (x_i, y_i) \|) \quad (3)$$

3. Results and Discussion

3.1. Weight Loss of Torrefied Biomass

Weight loss is an important parameter for optimizing the design and operation of a biomass torrefaction plant [6]. The mass yield (Table 1) under the different conditions of temperature and residence time was tested. In Table 1, which shows the average results obtained, the homogenous groups resulting from the analysis of the variance are indicated by superscripts. If two conditions have the same superscript, this means that no statistically significant differences were detected.

Table 1. Mass yield at the end of torrefaction versus the temperature and residence time tested. The superscripts ($^{A–H}$) show the homogeneous subsets indicated by the ANOVA test. If two conditions have the same superscript, this means that no statistically significant differences were detected.

Temperature (°C)	Residence Time (min)		
	0	30	120
200	97.48 ± 0.69 ^{A}	92.20 ± 0.65 ^{B}	89.48 ± 0.22 ^{C}
250	90.53 ± 0.54 ^{B,C}	79.87 ± 0.44 ^{E}	75.71 ± 0.82 ^{F}
275	84.57 ± 1.14 ^{D}	72.01 ± 0.55 ^{G}	66.49 ± 1.35 ^{G}
300	78.89 ± 0.84 ^{E}	62.38 ± 1.86 ^{H}	57.61 ± 0.80 ^{H}

It was observed that the mass yield of the biomass tested varied between 97.48% and 57.61%. At 200 °C, the weight loss was relatively low, reaching a maximum value of 10.52% with a residence time of 2 h (89.48% of mass yield). At this temperature (i.e., the first stage), the slight decay of the biomass weight was due to the drying procedure and the release of some light volatiles [27]. According to Chen et al., the weight loss between 250 and 300 °C could be caused by dehydration reactions via bond scission with the elimination of H_2O, carbonyl, and carboxyl group formation reactions with the elimination of CO and CO_2, and limited devolatilization and carbonization for the production of final tars and chars [17]. For this reason, at temperatures greater than 250 °C, the weight loss was higher, leading to drastic weight reductions at 300 °C, similar to those observed by Chin et al. [6]. At a residence time of 30 min, the mass started to decline dramatically from a temperature of 275 °C, similarly to the results obtained by Phanaphanic and Mani [33]. They used pine wood chips and logging residue chips as biomass with a residence time of 30 min and observed that the biomass weight decreased to only about one-half of its original value when the torrefaction temperature reached 300 °C.

Independently of the temperature, the weight loss increased with time. However, the effect of time was higher when the temperature was increased. At 200 °C, the difference in weight during the first 2 h was approximately 8%. This difference was significantly higher at 300 °C. At this temperature, the change in weight was 24.28%. At 250 and 275 °C, the change presented intermediate values (14.82% and 18.15%, respectively). In the same way, at a constant residence time, the weight loss also increased with temperature. The effect of the temperature on the weight loss was higher when the residence time was higher too. At 0 min, the difference between 200 and 300 °C was 18.59%, whilst at 2 h the difference obtained at the same temperatures was 31.87%. These results clearly show that the effect of the temperature is higher than the effect of the residence time, similarly to the observation of Nimlos et al. [34]. They studied sawdust torrefaction and reported that the torrefaction temperature has a more profound effect on the weight loss than the residence time. In fact, when comparing the weight loss between residence times of 0 and 30 min and between 30 and 120 min, it is clear than after the first 30 min, the weight loss is lower. At 200 °C, the difference in weight between 0 and 30 min was 5.38%, while between 30 and 120 min it was only 2.72%. This reduction is similar to that obtained under low temperature (200 °C) by Chin et al. [6] in a study about the optimization of torrefaction conditions in lignocellulosic biomass. This is more significant at higher temperatures (275 and 300 °C), at which the weight loss did not present statistically significant differences. Moreover, the ANOVA showed that the results obtained with a residence time of 120 min and a temperature of 200 °C were similar to those obtained at an initial time and 250 °C.

The ash content of the raw material was 11.51% ± 0.28%. This value is increased because of the presence of olive leaves [35]. After the torrefaction, a higher ash content of the torrefied mass was observed. The relative content of ash is noticeably increased in the torrefied biomass [22], which is related to the loss of mass of organic matter during torrefaction [36]. During torrefaction, the fixed compounds remain and the ratio of mass of fixed compounds to total mass increased in the same way as the weight loss. The ash content ranged between 11.85% ± 0.28% (200 °C and 0 min) and 19.95% ± 0.24% (300 °C and 120 min). This implied a maximum increase of 73.32%. However, this value is lower than that obtained by Pinchuai et al. [37] with other raw materials such as rice husk, sawdust, or bagasse (resulting in the ash content being increased by more than 100%). One problem inherent in biomass combustion stems from the ash generated in the process [38], so the increase of the ash content could be the main disadvantage for the use of torrefaction as a pretreatment.

3.2. Higher Heating Value (HHV)

The heating value of the biomass is an important property, as it determines its use in energy applications [20]. The higher heating value (HHV) of the raw material was 4933 ± 80 kcal/g. Considering its origin, this value is coherent with those obtained by other researchers investigating the use of olive waste, such as the value obtained by Zamorano et al. for leaves of olive trees [39]. When the biomass is torrefied, the HHV clearly intensifies, so compared to the HHV of raw material, the HHV increased by approximately 4.8 to 5.9 kcal/g under the different conditions tested. The average HHV for each condition tested is shown in Table 2.

It is observed that HHV varied from 4884.14 ± 36.45 to 5893 ± 68.83 kcal/g, representing an increase of 19.47% for 275 °C and 120 min. These results are similar to those obtained by Benavente and Fullana [4], who torrefied olive mill wastes at 300 °C and obtained a 13.4% rise in HHV. Chen et al. collected data on the increase of HHV after torrefaction in a review of the use of different raw materials and found that the rise in HHV reached almost 63% with a mixture of spruce, pine, and fir [19]. The increase in the HHV with increasing temperature was mainly due to a reduction of the low-energy bonds and an increase of high-energy bonds [12].

Table 2. Higher heating value of torrefaction versus the temperature and residence time tested. The superscripts ($^{A–F}$) show the homogeneous subsets indicated by the ANOVA test. If two conditions have the same superscript, this means that no statistically significant differences were detected.

Temperature (°C)	Residence Time (min)		
	0	30	120
200	5028.19 ± 102.76 ^{B}	5053.42 ± 33.27 ^{B}	5527.34 ± 68.97 ^{C,D}
250	5129.31 ± 59.75 ^{B}	5409.52 ± 26.03 ^{C}	5679.21 ± 69.72 ^{D,E}
275	4884.14 ± 36.45 ^{A}	5830.70 ± 44.31 ^{E,F}	5893.45 ± 68.83 ^{F}
300	4895.61 ± 57.65 ^{A,B}	5679.08 ± 66.34 ^{D,E}	5725.33 ± 21.23 ^{D,E,F}

Considering the ANOVA test, we can see the conditions under which the HHV was similar. At the initial time (0 min), the HHV was similar to that obtained with the raw material at all temperatures tested, slightly higher at 200 and 250 °C, and slightly lower at 275 and 300 °C. Considering this fact, operation at 275 or 300 °C and 0 min cannot be recommended. Similarly, at 200 °C, comparing the results obtained at residence times of 0 and 30 min, it was observed that the results did not present statistically significant differences, so performing the torrefaction at 0 min is recommended. Slightly higher results were detected at 250 °C and 30 min and at 200 °C and 2 h. In this case, a specific analysis considering other advantages and disadvantages must be done to choose the optimal conditions.

The highest HHV was detected at 275 °C and residence time of 120 min. From this temperature, the HHV decreased with this parameter. At 275 °C, the highest HHV was obtained at a residence time of 120 min (5893.45 ± 68.83 kcal/g). However, the ANOVA test showed that there were no statistically significant differences from the result achieved with a residence time of 30 min and temperature of 275 °C (5830.70 ± 44.31 kcal/g), so the optimal operating condition for the torrefaction of this biomass would be 30 min. At a temperature of 300 °C, the HHVs were 5679.08 ± 66.34 and 5725.33 ± 21.23 kcal/g at residence times of 30 and 120 min, respectively, presenting no statistically significant differences from the optimal condition. However, the weight loss at 300 °C is higher, as mentioned in the previous section, and due to this higher weight loss along with the lower HHV, this temperature is not recommended.

At a temperature of 200 °C, from 30 to 120 min, the HHV increased by 9.38%. However, the effect of the residence time on the HHV of torrefied biomass at high temperatures (275 and 300 °C) was significantly lower. As the ANOVA test showed, no statistically significant differences were detected between 30 and 120 min, which is coherent with the results of Arias et al. [19], who observed that the heating value yield of woody biomass remained practically constant from 30 min to 2 h of torrefaction.

3.3. Numerical Model

Several studies about the effect of the operational conditions in the torrefaction process have been carried out, e.g., on solid olive waste products to study changes in properties [15] and on almond shells using a response surface methodology to examine effects of torrefaction temperature and time on mass and energy yields of solid product [40]. This research presents as a novelty the development of a numerical model to predict the optimal conditions of temperature and residence time for the torrefaction of olive tree waste. The independent variables, x, y, of the numerical model were temperature and residence time, respectively. The dependent variables were the HHV and the mass yield loss. For these values, and according to the methodology explained in Section 2.6, the explicit expression of the approximant from 12 data points for the HHV, by using as basis function the multiquadric function, was obtained. Its graph is shown in Figure 2, where the axes on the plane represent temperature and residence time, and the vertical axis represents the approximated HHV. The graph for the mass yield approximant built from 16 points with translations of multiquadric functions is shown in Figure 3.

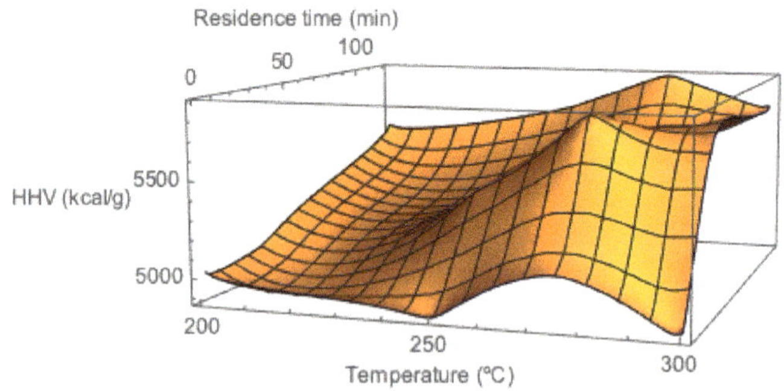

Figure 2. Graph for the higher heating value (HHV) approximant.

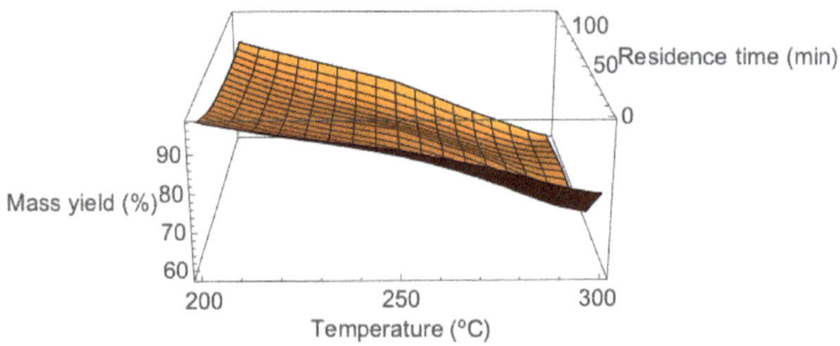

Figure 3. Graph for the mass yield approximant.

Once these functions were obtained, they were used to infer some facts of interest. To deduce them, the approximate values of residence time and temperature for which the maximum HHV is achieved, and the approximate values for which the strongest decrease of mass yield is reached (maximum weight loss), were computed.

With respect to the first goal, numerical techniques were used to determine that the maximum HHV on the interior domain of our variables was reached at 275.45 °C and 31.23 min. With respect to the second aim, and considering $s(x, y)$ the mass yield approximant, it is known that the value (x, y) for which the strongest decrease of the function $s(x, y)$ occurs is that one for which $-||\nabla s(x, y)||$ is minimized, where $\nabla s(x, y)$ is the gradient vector of $s(x, y)$. This coincides with the point for which $||\nabla s(x, y)||$ is maximized or, equivalently, to remove square roots, where $|| \nabla s(x, y) ||^2$ is maximized. The graph of $||\nabla s(x, y)||^2$ is shown in Figure 4, in which the horizontal axes represent the residence time and the temperature. By using numerical techniques again, the strongest decrease on the interior domain of our variables was achieved at 278.27 °C and 60.56 min (which corresponds to the values of temperature and residence time for which $|| \nabla s(x, y) ||^2$, represented on the vertical axis of Figure 4, reaches its maximum).

Our objective was to determine the values of residence time and temperature for which maximum HHV and small weight loss are achieved in order to maintain the maximum mass with the highest heating value. Taking into account the values cited in the two previous paragraphs, it was determined that fixing the smallest temperature, that is, 275.45 °C, the weight loss between 31.23 and 60.56 min is less than three percent (this can be easily seen in Figure 5, in which the graph for the weight loss approximant is focused on a range of residence time between 25 and 65 min and a range of temperature between 265 °C and 280 °C). As a result of this, the conclusion is that an optimal choice for our proposals would be approximately 275 °C and 30 min. These conditions are similar to those obtained by Chin et al. [6] for oil palm biomass and fast-growing species available in Malaysia, *Acacia* spp.

(260 °C for 30 min) and *Macaranga* spp. (280 °C for 45 min), in a study about the optimization of torrefaction conditions.

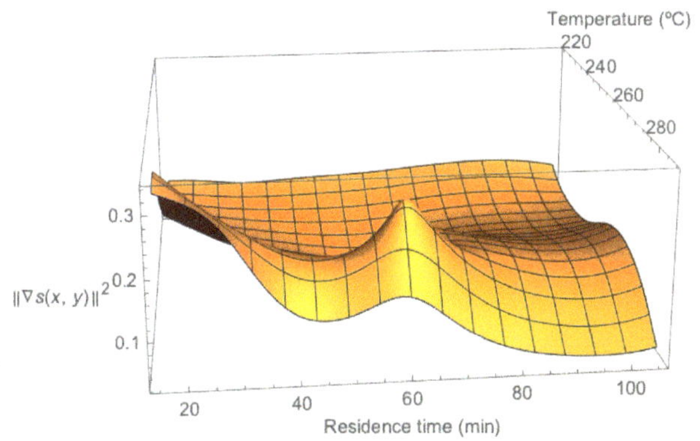

Figure 4. Graph of $\| \nabla s(x, y) \|^2$.

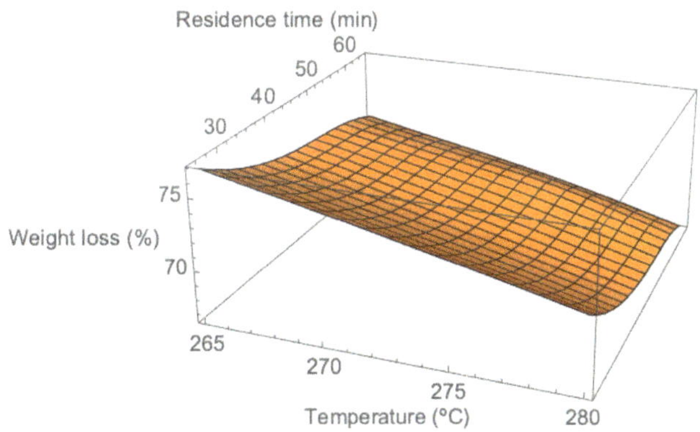

Figure 5. Graph for the weight loss approximant.

4. Conclusions

From the results obtained for the torrefaction of olive tree waste at four different temperatures (200, 250, 275, and 300 °C) and four different residence times (0, 30, 60, and 120 min), the following conclusions were drawn:

(1) The mass yield of the tested biomass varied between 97.48% and 57.61%, decreasing with the residence time and temperature. However, the ash content of the torrefied biomass increased with respect to the raw material up to a maximum of 73.32%, so the problem inherent in biomass combustion of ash could be increased.

(2) The HHV of the torrefied biomass varied from 4884.14 ± 36.45 to 5893 ± 68.83 kcal/g. In the best case (275 °C and 120 min), a 19.47% increase in the HHV was achieved. The results of the mathematical model showed that the optimal conditions for torrefaction of olive tree waste are approximately 275 °C and 30 min.

For all these reasons, the use of torrefaction for this olive tree biomass could be a reliable pretreatment when operating under the optimal conditions.

Author Contributions: Conceptualization, J.M.-P.; laboratory methodology, J.M.-P.; validation J.J. and M.L.R.; formal analysis M.Z., J.M.-P., M.L.R. and J.J.; writing—original draft preparation, J.M.-P.; writing—review and editing, M.Z.; supervision, M.Z. All authors have read and agreed to the published version of the manuscript.

Funding: This research received no external funding.

Acknowledgments: The authors appreciate the support of the research groups TEP-968 (Technologies for Circular Economy) and FQM191 of the University of Granada, and EIFQM8 and FQM178 of the University of Jaén.

Conflicts of Interest: The authors declare no conflict of interest.

References

1. Volpe, M.; Wüst, D.; Merzari, F.; Lucian, M.; Andreottola, G.; Kruse, A.; Fiori, L. One stage olive mill waste streams valorisation via hydrothermal carbonization. *Waste Manag.* **2018**, *80*, 224–234. [CrossRef] [PubMed]
2. Lenis, Y.A.; Agudelo, A.; Pérez, J.F. Analysis of statistical repeatability of a fixed bed downdraft biomass gasification facility. *Appl. Therm. Eng.* **2013**, *51*, 1006–1016. [CrossRef]
3. Ramos-Carmona, S.; Martínez, J.D.; Pérez, J.F. Torrefaction of patula pine under air conditions: A chemical and structural characterization. *Ind. Crops Prod.* **2018**, *118*, 302–310. [CrossRef]
4. Benavente, V.; Fullana, A. Torrefaction of olive mill waste. *Biomass Bioenergy* **2015**, *73*, 186–194. [CrossRef]
5. Ibrahim, R.H.H.; Darvell, L.I.; Jones, J.M.; Williams, A. Physicochemical characterization of torrefied biomass. *J. Anal. Appl. Pyrolysis* **2013**, *103*, 21–30. [CrossRef]
6. Chin, K.L.; H'ng, P.S.; Go, W.Z.; Lim, T.W.; Maminski, M.; Paridah, M.T. Optimization of torrefaction conditions for high energy density solid biofuel from oil palm biomass and fast growing species available in Malaysia. *Ind. Crops Prod.* **2013**, *49*, 768–774. [CrossRef]
7. Berndes, G.; Hoogwijk, M.; van den Broek, R. The contribution of biomass in the future global energy supply: A review of 17 studies. *Biomass Bioenergy* **2003**, *25*, 1–28. [CrossRef]
8. Zheng, A.; Zhao, Z.; Chang, S.; Huang, Z.; He, F.; Li, H. Effect of torrefaction temperature on product distribution from two-staged pyrolysis of biomass. *Energy Fuels* **2012**, *26*, 2968–2974. [CrossRef]
9. Da Silva, C.M.S.; Vital, B.R.; Carneiro, A.D.C.O.; Costa, E.V.; De Magalhães, M.A.; Trugilho, P.F. Structural and compositional changes in eucalyptus wood chips subjected to dry torrefaction. *Ind. Crops Prod.* **2017**, *109*, 598–602. [CrossRef]
10. Matos, M.; Barreiro, M.F.; Gandini, A. Olive stone as a renewable source of biopolyols. *Ind. Crops Prod.* **2010**, *32*, 7–12. [CrossRef]
11. Nunes, L.J.R.; Loureiro, L.M.E.F.; Sá, L.C.R.; Silva, H.F. Thermochemical Conversion of Olive Oil Industry Waste: Circular Economy through Energy Recovery. *Recycling* **2020**, *5*, 12. [CrossRef]
12. Garcia-Maraver, A.; Zamorano, M.; Ramos-Ridao, A.; Díaz, L.F. Analysis of olive grove residual biomass potential for electric and thermal energy generation in Andalusia (Spain). *Renew. Sustain. Energy Rev.* **2012**, *16*, 745–751. [CrossRef]
13. Mei, Y.; Che, Q.; Yang, Q.; Draper, C.; Yang, H.; Zhang, S.; Chen, H. Torrefaction of different parts from a corn stalk and its effect on the characterization of products. *Ind. Crops Prod.* **2016**, *92*, 26–33. [CrossRef]
14. Li, M.F.; Li, X.; Bian, J.; Xu, J.K.; Yang, S.; Sun, R.C. Influence of temperature on bamboo torrefaction under carbon dioxide atmosphere. *Ind. Crops Prod.* **2015**, *76*, 149–157. [CrossRef]
15. Martín-Lara, M.; Ronda, A.; Zamora, M.C.; Calero, M. Torrefaction of olive tree pruning: Effect of operating conditions on solid product properties. *Fuel* **2017**, *202*, 109–117. [CrossRef]
16. Prins, M.J.; Ptasinki, K.J.; Jansen, F.J.J.G. Torrefaction of wood Part 2. Analysis of products. *J. Anal. Appl. Pyrolysis* **2006**, *77*, 35–40. [CrossRef]
17. Chen, W.-H.; Peng, J.; Bi, X.T. A state-of-the-art review of biomass torrefaction, densification and applications. *Renew. Sustain. Energy Rev.* **2015**, *44*, 847–866. [CrossRef]
18. Yan, W.; Acharjee, T.C.; Coronella, C.J.; Vasquez, V.R. Vásquez, Thermal pretreatment of lignocellulosic biomass. *Environ. Prog. Sustain. Energy* **2009**, *28*, 435–440. [CrossRef]
19. Arias, B.; Pevida, C.; Fermoso, J.; Plaza, M.; Rubiera, F.; Pis, J.J. Influence of torrefaction on the grindability and reactivity of woody biomass. *Fuel Process. Technol.* **2008**, *89*, 169–175. [CrossRef]
20. Chen, W.-H.; Kuo, P.-C. Torrefaction and co-torrefaction characterization of hemicellulose, cellulose and lignin as well as torrefaction of some basic constituents in biomass. *Energy* **2011**, *36*, 803–811. [CrossRef]

21. Park, S.-W.; Jang, C.-H.; Baek, K.-R.; Yang, J.-K. Torrefaction and low-temperature carbonization of woody biomass: Evaluation of fuel characteristics of the products. *Energy* **2012**, *45*, 676–685. [CrossRef]
22. Van Der Stelt, M.J.C.; Gerhauser, H.; Kiel, J.H.A.; Ptasinski, K.J. Biomass upgrading by torrefaction for the production of biofuels: A review. *Biomass Bioenergy* **2011**, *35*, 3748–3762. [CrossRef]
23. Chen, W.-H.; Lu, K.-M.; Liu, S.-H.; Tsai, C.-M.; Lee, W.-J.; Lin, T.-C. Biomass torrefaction characteristics in inert and oxidative atmospheres at various superficial velocities. *Bioresour. Technol.* **2013**, *146*, 152–160. [CrossRef]
24. Bach, Q.-V.; Skreiberg, Ø. Upgrading biomass fuels via wet torrefaction: A review and comparison with dry torrefaction. *Renew. Sustain. Energy Rev.* **2016**, *54*, 665–677. [CrossRef]
25. Bach, Q.-V.; Tran, K.-Q.; Skreiberg, Ø.; Trinh, T.T. Effects of wet torrefaction on pyrolysis of woody biomass fuels. *Energy* **2015**, *88*, 443–456. [CrossRef]
26. Wannapeera, J.; Worasuwannarak, N. Upgrading of woody biomass by torrefaction under pressure. *J. Anal. Appl. Pyrolysis* **2012**, *96*, 173–180. [CrossRef]
27. Simonec, M.; Goricanec, D.; Urbancl, D. Impact of torrefaction on biomass properties depending on temperature and operation time. *Sci. Total Environ.* **2020**, *740*, 140086. [CrossRef]
28. Harun, N.H.H.M.; Samad, N.A.F.A.; Saleh, S. Development of kinetics model for torrefaction of empty fruit bunch from palm oil waste. *Energy Procedia* **2017**, *105*, 744–749. [CrossRef]
29. Nakason, K.; Pathomrotsakun, J.; Kraithong, W.; Khemthong, P.; Panyapinyopol, B. Torrefaction of Agricultural Wastes: Influence of Lignocellulosic Types and Treatment Temperature on Fuel Properties of Biochar. *Int. Energy. J.* **2019**, *19*, 253–266.
30. Burden, A.; Burden, R.; Faires, J. *Numerical Analysis*, 10th ed.; Cengage: New York, NY, USA, 2016.
31. Buhmann, M.D. *Radial Basis functions*; Cambridge University Press: Cambridge, UK, 2003. [CrossRef]
32. Fasshauer, G.E. *Meshfree Approximation Methods with Matlab*; World Scientific Publishing: Singapore, 2007. [CrossRef]
33. Phanaphanich, M.; Mani, S. Impact of torrefaction on the grindability and fuel characteristics of forest biomass. *Bioresour. Technol.* **2011**, *102*, 1246–1253. [CrossRef]
34. Nimlos, M.R.; Brooking, E.; Michael, J. Biomass torrefaction studies with a molecular beam mass spectrometer. *Am. Chem. Soc. Div. Fuel Chem.* **2003**, *48*, 590–591.
35. García-Maraver, A.; Terron, L.C.; Ramos-Ridao, A.; Zamorano, M. Effects of mineral contamination on the ash content of olive tree residual biomass. *Biosyst. Eng.* **2014**, *118*, 167–173. [CrossRef]
36. Bridgeman, T.G.; Jones, J.M.; Williams, A.; Waldron, D.J. An investigation of the grindability of two torrefied energy crops. *Fuel* **2010**, *89*, 3911–3918. [CrossRef]
37. Pinchuai, A.; Dutta, A.; Basu, P. Torrefaction of Agriculture Residue to Enhance Combustible Properties. *Energy Fuels* **2010**, *24*, 4638–4645. [CrossRef]
38. García-Maraver, A.; Mata-Sánchez, J.; Carpio, M.; Pérez-Jiménez, J.A. Critical review of predictive coefficients for biomass ash deposition tendency. *J. Energy Inst.* **2017**, *90*, 214–228. [CrossRef]
39. Zamorano, M.; Popov, V.; Rodríguez, M.L.; García-Maraver, A. A comparative study of quality properties of pelletized agricultural and forestry lopping residues. *Renew. Energy* **2011**, *36*, 3133–3140. [CrossRef]
40. Chiou, B.-S.; Valenzuela-Medina, D.; Bilbao-Sainz, C.; Klamczynski, A.P.; Avena-Bustillos, R.J.; Milczarek, R.R.; Du, W.-X.; Glenn, G.M.; Orts, W.J. Torrefaction of almond shells: Effects of torrefaction conditions on properties of solid and condensate products. *Ind. Crops Prod.* **2016**, *86*, 40–48. [CrossRef]

sustainability

MDPI

Article

Analysis of District Heating and Cooling Energy Systems in Spain: Resources, Technology and Management

Beatriz María Paredes-Sánchez [1,*], José Pablo Paredes [1], Natalia Caparrini [2] and Elena Rivo-López [3]

1. School of Mining, Energy and Materials Engineering of Oviedo (EIMEM), University of Oviedo, 33004 Oviedo, Spain; paredespablo@uniovi.es
2. Department of Natural Resources and Environmental Engineering, University of Vigo, 36310 Vigo, Spain; nataliac@uvigo.es
3. Department of Business Organization and Marketing, Faculty of Business and Tourism, University of Vigo, 32004 Ourense, Spain; rivo@uvigo.es

* Correspondence: UO19070@uniovi.es; Tel.: +34-985-10-43-05

Abstract: District heating and cooling (DHC) systems play an important role under the new European Union (EU) energy transition strategy. Thermal energy networks are helping to stimulate the development of alternative technologies based on a broad range of renewable energy sources. The present study analysed the current situation of DHC systems in Spain and provides an overview of the challenges and future opportunities that their use will entail. Its objective is to assess thermal energy conversion and management from a holistic perspective, including a study of existing energy infrastructures. The focus of this study lies on Spain given the country's abundance of natural resources such as renewable energy sources including solar energy, biomass and geothermal energy, among others, as well as its strategic location on the map of the EU. Based on the analysis of the three factors for energy conversion in a district heating system, namely resources, technology, and management, the methodology provided an assessment of the different factors involved in running a DHC system. The results show an estimated total production for DHC networks of 1448 MW_{th}, of which 72% is supplied purely by renewable energy sources.

Keywords: energy conversion; energy management; technology; thermal system

Citation: Paredes-Sánchez, B.M.; Paredes, J.P.; Caparrini, N.; Rivo-López, E. Analysis of District Heating and Cooling Energy Systems in Spain: Resources, Technology and Management. *Sustainability* **2021**, *13*, 5442. https://doi.org/10.3390/su13105442

Academic Editor: Tomonobu Senjyu

Received: 15 April 2021
Accepted: 10 May 2021
Published: 13 May 2021

1. Introduction

Energy is a basic need for society and economic growth [1]. Accomplishing thorough decarbonisation will require full implementation of climate targets in all sectors. In order to achieve these targets in the energy sector, in particular, renewable energies are promoted as alternatives to fossil fuels. The first step in studying their viability is to analyse the factors related to their use and management systems. However, the COVID-19 pandemic has spurred on the energy transition and has turned renewable energies into a sector with great business potential, to the point that fossil fuel-based traditional energy companies have taken a green turn in their strategies.

Power systems must be able to meet the current demand for thermal and electrical energy using available technology, i.e., drawing on existing technology to achieve a more sustainable energy performance [2].

Currently, residential and commercial buildings account for almost one-third of global greenhouse gas (GHG) emissions. Indeed, recent studies estimate that global energy consumption and GHG emissions will rise by approximately 30% by 2040, which means that technological development will play a key role in addressing environmental issues resulting from this technological challenge [3]. Therefore, reducing the use of fossil fuels in the energy production process will make a significant contribution towards meeting the set global targets for reducing the use of fossil fuels [4].

District heating and cooling (DHC) in buildings and industry accounts for half of the European Union's (EU) energy consumption, and 75% of it is generated from fossil fuels. Spain ranked 25th in the EU in terms of household final energy consumption per capita in 2016, where the country consumed 324 kg of oil equivalent of electricity and heat per capita, excluding transport. Foreign energy dependence stood at 73.9% in 2017, which is two and a half points above the previous year [5].

Against this background, the EU presented the "Green Deal" at the end of 2019, which put forward a new strategy towards a thriving and fair society founded on a resource-efficient economy aiming to achieve climate neutrality by 2050. This meant an increase in ambition that should be reflected in an upwards revision of the current 40% emission reduction target set for 2030. The Spanish government, for its part, is working on a "Climate Change and Energy Transition Law" and has presented a draft of the "Integrated National Energy and Climate Plan" (PNIEC), with ambitious objectives for a practically decarbonised economy by 2050.

In this sense, energy management strategies in polygeneration systems that integrate multiunit connections involving different natural resources as energy sources will lead to better and more efficient systems [6].

On the other hand, projects aimed at improving energy efficiency in individual buildings cannot offset the increased energy demand created by new buildings. These projects are costly and time-consuming, albeit necessary, when applied to existing buildings.

A circular economy aims to keep valuable resources for as long as possible while restricting waste generation to a minimum. A circular economy should lead to lower energy consumption and carbon dioxide emissions from local to global levels.

It is in this context of energy transition and circular economy that district heating and cooling systems could make an important contribution to the construction sector by improving the energy conditions of buildings while meeting decarbonisation targets.

District heating (DH) networks are designed for collective use, which requires a large surface area to capture solar energy and allow for the use of a combination of alternative energy fuel sources (both fossil and renewable) in existing systems.

The term DH appeared in Europe at the beginning of the 20th century. DH is a system for distributing heat generated in a centralized location through a system of insulated pipes for residential and commercial heating requirements such as space heating and water heating. Now, newer configurations, known as district heating and cooling (DHC) systems, could meet energy demands for both heating and cooling [7].

Urban heating and cooling systems are especially common in Scandinavian, Baltic, and Eastern European countries, many of which have a long history of using them, and new thermal systems can often be adapted to existing infrastructure. At present, Spain is also making important strides in the implementation of these types of energy systems [8].

Currently, the rapid growth of DHC systems allows more efficient use of local renewable resources within the European energy market [9]. Additionally, DHC systems often use local fuels and resources, which would otherwise be wasted, in order to meet local heating energy demands through local distribution networks. Traditionally, heating networks were most commonly powered by residual thermal energy and/or fossil fuel combustion. However, over the past few decades, DHC networks have begun incorporating several alternative renewable energy sources. Likewise, they are incorporating more recycled and renewable heat, which has become the main focus on urban heating systems today [10]. District heating networks can be fuelled by some heat generation sources, including combustion plants (based either on fossil fuels or biomass), CHP plants (combined heat and power), or renewable energy-based plants (e.g., biomass, solar, or geothermal). A multiple-heat-source combination solution is beneficial, particularly for large district heating schemes [11]. Solar energy plays a relevant role in thermal applications, e.g., the solar collector technology for buildings. Boiler stations are specifically devoted to generating thermal energy, which is produced by combustion of fossil fuels (i.e., natural gas, heating oil, or coal), or renewable fuels (i.e., biomass or solid waste). Unlike boiler systems, which are specifically dedicated to producing thermal energy, CHP systems deliver thermal energy as a product of electricity generation. A CHP system can achieve more than 80% energy efficiency [12]. Typical boiler efficiencies in energy systems range from approximately 90% with the best solid biomass-fuelled boilers to 95% with natural gas-fuelled boilers. However, heat pump-based systems enabling heat recovery from the ground (i.e., geothermal heat pump systems) and alternatively from other low-grade heat sources, typically have a coefficient of performance (COP) of around 4 [12–14]. Additionally, DH systems running on waste heat provide a way to efficiently manage fuel for space heating, which may be originally sourced from fossil fuels. Energy storage has become an important aspect of DH networks. Thermal energy storage (TES) is a type of technology used to store thermal energy by heating or cooling a storage medium. There are two main types of thermal energy storage, thermal (sensible heat and latent heat) and chemical [12]. Such stored energy can be further used for heating and cooling purposes. TES efficiency values can exceed 70% [15]. Heat, cooling, and electricity production (trigeneration systems) allows CHP technology to be integrated with heat pumps. Trigeneration technologies coupled with fuel cells are instrumental in the use of emerging alternative energy sources such as hydrogen. Micro CHP fuel cells, direct flame combustion boilers, catalytic boilers, and gas-fired heat pumps could all be fuelled with hydrogen. An array of larger thermal systems and industry devices running on natural gas also could be redesigned to use hydrogen [16]. In this sense, residual biomass as a renewable resource has been used in trigeneration for high-efficiency thermal blanket heating applications, with the integration of solid oxide fuel cells (SOFC) and gasifier [17,18].

In Spain, the consumption of renewable thermal energy has risen to 50,732 GWh. Biomass accounted for 91.95% of this total, followed by thermal solar (6.73%), biogas (0.88%), and geothermal (0.45%) [19]. Given its strategic importance, it is fundamental that all Spanish bioeconomy strategy policies establish the development of bioenergy as a key priority in the future [20].

Silva et al. [21] show open challenges where the smart city concept is still evolving throughout the globe due to economic and technological barriers. Several case studies have already demonstrated the importance of DHC networks [22–25]. The majority of these studies focused on one aspect or domain within DH/DHC systems and attempted to connect the entire system according to the type of each resource, technology, or energy management strategy. Several authors have studied the existing heating networks from different perspectives. Mazhar et al. [26] analysed the progress that has been made in technology and proactive research methods to minimise carbon emissions within the heating industry. Vandermeulen et al. [27] argued the need to develop more advanced control systems to improve overall energy management. Lund et al. [28] demonstrated the strong technical and economic potential of these systems and their ability to provide a viable source of heating and cooling for the future. Akhtari et al. [4] and Lake et al. [29] highlighted the need for future network heating system studies that would include factors such as resources, technology, and energy management.

This document aims to fill the existing gaps in the literature on energy sources and implementation of district heating systems, thus providing a framework for research into the DHC system that is in line with the principles of sustainable development. To this end, the three energy conversion factors—resources, technology, and management—were studied, applying them as an example to district heating systems in Spain from a time transition perspective, to achieve more widespread implementation of renewable energy sources and more efficient energy conversion in the future.

This work studies the resources, technology, and energy management of DHC systems from a time perspective of progressive implementation in Spain and is therefore intended as a useful tool to be used for similar processes worldwide.

The novelty of this work lies with the effective identification of actions and limitations in the DHC systems. In this sense, it combines technical, economic, and environmental data regarding the resources, the available technology and the energy management of these systems. Furthermore, it aims to provide a framework for research into the DHC system that is in harmony with the principles of sustainable development: need, equity, generation transition and global environmentalism.

The present study is organized as follows: Section 1 includes the introduction, aims, and gaps of knowledge in the sustainability context; Section 2 explains the analytical methodology applied to the different elements involved in the energy conversion process; Section 3 presents the results within the current framework of available energy resources, technology, and management strategies, and comprises the core of the work; Section 4 examines both the opportunities, challenges facing the industry at present andprovides the final observations; and Section 5 shows the conclusions.

2. Materials and Methods

The work is intended to provide an analysis of the current use of the DHC system to identify potential technological developments and help expand the use of multilateral systems of thermal energy management in Spain following European policies and regulations. More specifically, this study includes an analysis of relevant information and studies published between 2010 and 2019.

The used methodology was based on three phases of energy conversion: resources, technology, and management [30]. Factors such as energy, the environment, and management were analysed under the energy context [22,30]. An analysis of the driving forces yielded data on the actual actions and limitations in DHC systems with the available information of the main existing databases for DHC systems.

Databases and inventories of both public and private organizations of reference with jurisdiction in DH/DHC systems were searched. The resulting multi-objective methodology was based on three specific phases of energy conversion in DHC systems. The first step consisted of the collection of data related to the energy sources used within the studied territory (Phase 1: Resource). Two challenges arose during the process of evaluating the energy potential in a conversion analysis: discrepancies in statistical data and the difficulty involved in calculating the real energy potential [30]. The second step was to analyse the available existing technology (Phase II: Technology). Finally, the systems were examined from an energy management perspective within the current regulatory framework, and their prospects were outlined (Phase III: Energy Management). The research framework is shown in Figure 1.

Furthermore, a critical analysis of the available scientific literature was conducted to fill the existing knowledge gaps to understand the relationship among resources, technology, and the energy conversion management process in these systems. Figure 1 shows a graph of the methodology used.

The core of the database analysis is rooted in sources provided by Spanish energy institutions (Table 1).

Table 1. Databases of analysis and input data.

Organization	Database Resource	Reference
ADHAC (Association of District Heating and Cooling Companies)	Industrial sector and technology	[31]
APPA (Association of Renewable Energy Producers)	Industrial sector	[19]
EurObserv'ER	Geographical and social parameters, production, and technology	[8]
IDAE (Institute for Energy Diversification and Savings)	Resource characteristics, regulation, production, and financial Support	[32]
Spanish Biomass Technology Platform	Industrial sector	[20]
RHC (European Technology and Innovation Platform on Renewable Heating and Cooling)	Production parameters	[33]

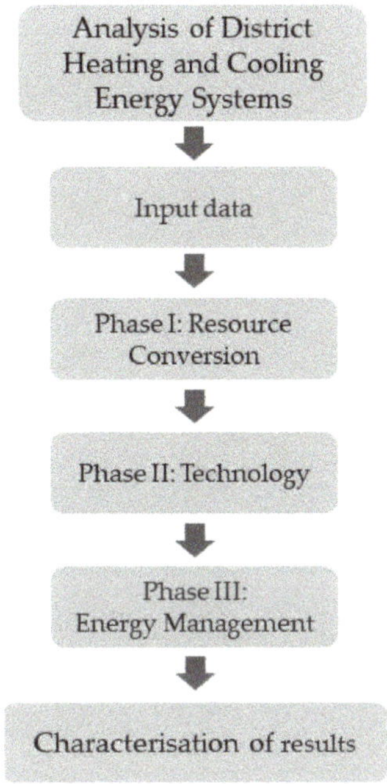

Figure 1. Methodological structure.

Literature analysis was updated with the present study so as to provide the latest developments, including driving force analysis based on the study of seasonal data from a range of sources of relevance to the study and future implementation of DHC systems, as contained in this document.

The respective zones of study for the methodology (Figure 1) were subdivided by region and assigned individual area codes (Table 2).

Table 2. Spanish zones by area code.

Region	Area Code
Andalusia	1
Aragon	2
Principality de Asturias	3
Balearic Islands	4
Canary Islands	5
Cantabria	6
Castile–La Mancha	7
Castile and Leon	8
Catalonia	9
Community of Valencia	10
Extremadura	11
Galicia	12
La Rioja	13
Community of Madrid	14
Region of Murcia	15
Autonomous Community of Navarre	16
Basque Country	17
Ceuta	18
Melilla	19

Figure 2 shows the distribution of the area codes in the study area of Spain.

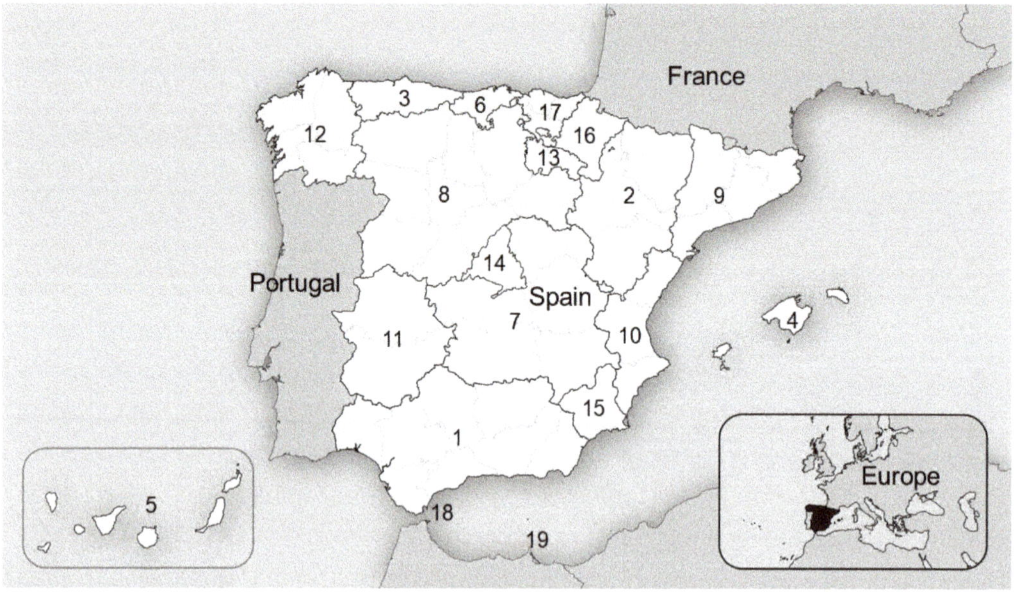

Figure 2. Distribution of the area codes in Spain.

The results of this study allow for the identification of several challenges related to the availability, management, and environmental impact of energy conversion in our society and offer suggestions to improve future research into DHC systems.

3. Results

District heating networks help improve energy efficiency in the service and construction sectors by offering more efficient climate control and, in doing so, help reduce overall energy demand with renewable energy. Energy demand is a key factor in the calibration of building climate control systems [34]. District systems allow for greater use of renewable energy sources and provide more efficient energy production, thereby reducing regional carbon emissions.

3.1. Phase I: Resource Conversion

3.1.1. Non-Renewable Sources

The main nonconventional or alternative energy sources available comprise those of renewable, reusable, or residual nature. Several different fuels are commonly used in residential buildings; natural gas stands out among conventional fuels as producing the least hazardous levels of emissions [35]. Conventional fossil fuel sources such as propane, butane, diesel, and coal are generally not used in DHC systems. largely because of the transport and storage difficulties involved as well as their higher levels of hazardous emissions.

The heat generated from these systems is used for both heating and domestic hot water (DHW) and is capable of supplying hot water ranging in temperature from 45 to 110 °C for either heating or steam-based systems. Hot water can be generated by heat pumps, boilers, CHP systems, or the use of residual energy sources (e.g., steam from a waste revaluation plant or smoke from industrial production). Although cooling is generally used for air conditioning purposes, it can also be used in either industrial processes or condensation circuits and is supplied through cold water, generally at around 5 °C. Steam can also generate heating and DHW through the use of steam/water exchangers or serve industrial purposes as a heat carrier fluid (at different pressures and temperatures, although it is most often superheated). While it is most often generated using compressors, it can also be harnessed from nearby sources of residual industrial energy.

Industrial cooling is usually generated using condensation circuits, compressors, or cold stores ranging from 0 to 7 °C and is supplied in a glycol/water mixture (at around −10 °C) or liquid carbon dioxide or ammonia. Generally, it is generated using either compressors or residual industrial energy sources.

3.1.2. Renewable Sources

There are also several highly efficient technological solutions that are compatible with the use of biomass, such as biogas, geothermal and thermal solar energies, high-efficiency combined heat and power, and residual heat from thermal energy power plants, waste management valuation plants, and industrial production (cement, glass, iron and steel, and aluminium as well as metalworking and forging). Besides its environmental advantages, such as the reduction of CO_2 emissions, biomass is the most common source of primary energy in heat networks because it has other advantages in line with savings and sustainable development. As it is indigenous and therefore uses resources from the environment in which it is consumed, it is not affected by the volatility of the fossil fuel market and presents societal advantages related to the creation of new economic activities in the environment and the improvement of incomes. Biomass can originate from different sources, including forestry and lumber industry residues, or it can appear in the form of biogas, which is a residual fuel source obtained from processing waste from landfills, sewage treatment plants, or urban/animal waste treatment plants [36].

The use of biomass yields some clear benefits, including greater symbiosis between a variety of industries and local communities and a wealth of social benefits (employment opportunities, urban heating, waste removal) generated in the production process [37]. Hagos et al. [38] discussed the importance of urban heating networks and individual and central bioheating systems in high energy demand areas to highlight the potential long-term benefits of bioenergy over conventional systems (2009–2030).

Thus, biomass will be a core element in the progress of Europe's bioeconomy and is one of the principal challenges related to both climate change and the energy transition process currently facing the EU.

There is a growing need to better understand and assess several of the key factors in global demand for bioenergy, including how much available biomass can be transported, how much is used and to what end, how it flows within the economy, and how a greater dependence on natural resources can be reconciled with meeting environmental, economic, and social sustainability standards at a European and global level. Moreover, new energy systems are constantly being implemented. Lausselet et al. [39] demonstrated the need for a circular economy when dealing with the management and use of these resources. Wood-based biomass is an efficient source of thermal energy [40]. A 51 kt pellet production is equivalent to 32 MW_{th} in thermal systems [41,42] (Figure 3).

A building's average primary energy consumption depends on the climate area. This can reach up to 282 kWh/(m^2 year) in northern Spain [43]. The heating consumption of an average Spanish household can reach up to 4700 kWh/year. This energy is basically supplied by electricity, natural gas, and diesel fuel [44].

Solar-powered urban DH systems have a long history of use in Europe. Sweden was the first country to develop this type of system, and a number of other European countries (including Denmark and Austria) subsequently recognised their enormous potential and fast-tracked the development of their thermal solar heating systems [45–47]. For thermal solar DH systems, the output range varies depending on the technology used in 1000 kW_{th} systems, which operate for 1500 h annually. The upper limit corresponds to installations with a concentrated collection tube, whereas the lower limit corresponds to installations with a coated/covered flat collector.

Figure 3. Renewable sources: forest biomass.

The use of geothermal energy, whether direct or through heat pumps, is an example of another highly efficient energy application. Centralised systems allow much greater output levels and higher efficiency compared to individual systems. The basic energy services commonly provided in DH systems include heating, cooling, steam supply, and industrial cooling. Geothermal energy systems are also used, which operate with an underground renewable energy source. Of these systems, those that work the equivalent of 3500 h provide a wide range of overall performance. The typical lower and upper limits of thermal power are 500 kW_{th} and 10,000 kW_{th}, respectively.

3.2. *Phase II: Technology*

3.2.1. Performance Principles

At present, Spain uses a wide range of different technologies that pose several challenges in terms of energy management. DH/DHC systems often vary depending on local energy policies, energy security, level of economic development, access to emerging and innovative technology, fuel dependency, regulations, climate, and other local conditions. For example, in the European territory, Poland uses geothermal heating technology even though current economic research shows that it is more expensive than coal and has a much lower calorific value than biomass, natural gas, and fuel oil [48].

Heating source flexibility is one essential element that all of these systems share, as any number of different centralised and decentralised heat sources can be used to provide dependable and flexible operating conditions using basic control strategies.

The main final objective for urban heating companies is to ensure that clients receive the lowest possible price for thermal energy, which requires a holistic approach considering that there are a growing number of heating and cooling options available.

District networks can also integrate renewable energy sources by using heat pumps, biomass and thermal solar energy, residual heating, and municipal waste. Depending on the location and the needs of any given zone, the same system can provide both heating during winter and cooling during the summer months using the same energy source year-round. Therefore, DH/DHC systems differ greatly in terms of energy management and environmental impact.

Thermal systems are characterised based on different factors: heat transfer fluid (e.g., air or water), transported thermal energy (e.g., cold, heat or both) or type of thermal resources (e.g., renewable or non-renewable). Energy efficiency is thus a key performance indicator of energy system [26] (Table 3).

Table 3. Summary of energy technology [26,49–56].

Source	Description	Performance Indicators	Barrier Parameters
Biomass	Uses wood-based input material to produce thermal energy. The oldest source for heating has been wood chips and wood pellets.	It has high thermal efficiency in energy systems, reaching a thermal efficiency of around 80–90%. Today, large-scale production of biofuels for DH grids allows for both economic and environmental benefits, enabling the energy supply to be managed, since it is a source of energy in the form of fuel.	There is low availability of biomass. A barrier to its mass commercialisation is its cost and the lack of adequate infrastructure. However, a versatile range of energy sources allows selecting the best fit for each set of applications to achieve the best performance.
Geothermal energy	This is the oldest and most mature of all DHC technologies. Most research seeks to improve energy efficiency and use geothermal heat in hybridisation with other energy sources.	It is built on sites above large geothermal or mining sources. Heat pumps increase the overall energy efficiency in heating and cooling performance. It provides low-cost heating and cooling by using heat pump technology in the DH system (thermal conversion efficiency above 60%) with a typical COP value of 4 in the case of heat pumps.	Geologically limited. The low efficiency of geothermal heat sources is partly because they are indirectly used for heating, given the potential for contamination in central heating systems in buildings.
Fossil fuel/waste heat	An old, mature heating generation mechanism. It burns coal, oil, or natural gas to provide thermal energy. The technology to implement this idea is available and is, in fact, widely used.	Energy infrastructure is often already running, thus reducing fuel transport-related costs. It is highly thermally efficient (85–95%). Fossil fuel waste energy could contribute to its implementation.	A non-renewable energy source producing high GHG emissions. Clean combustion and efficient waste management strategies hold the key to addressing this problem.
Solar	A mature technology, with most research aiming to both improve efficiency and incorporate heat storage. Sunlight and solar collectors are used to provide high-temperature water for thermal energy purposes.	High energy source availability with thermal conversion by both passive and active systems (thermal efficiency 30–80%). It is a low-grade heat source. Efficiency improvements could boost thermal output, particularly in regions with low solar irradiation.	Geographical assessment and proper planning are necessary. As solar thermal energy is unpredictable, it is not a reliable option in the absence of large-scale TES.

As a complement to the energy system technologies in Table 3, TES can operate as heat sinks at off-peak times and as peak demand heat sources in boiler, CHP, or trigeneration systems. TES systems for residential buildings ranges could overcome barriers such as energy supply variability from unpredictable and fluctuating renewable heat sources [57,58], and thus are expected to become integrated into DH/DHC network with efficiency over 70% [15]. Moreover, a variety of larger thermal systems and industry devices that use natural gas for thermal purposes could also be redesigned to use hydrogen for thermal purposes to develop DH/DHC networks [16].

3.2.2. Energy System Network Design

Energy systems are designed to meet the entire demand for heating, cooling, and DHW. Energy systems depend on several factors such as the fuel used, the technology, and the chosen location (Table 3).

An ideal scenario in urban areas would be either to harness the residual thermal energy from existing plants in operation or to create new ones that can harness either the residual thermal energy from the production of electricity or any residual fuel.

The current trend, however, is for power plants to be located outside the urban centre. Boilers or cogeneration equipment can be used to generate thermal energy in the form of heat, whether it be engines or turbines. Each technology can be used in combination with any of the various available energy sources, thereby yielding varying levels of emissions, with higher emissions from fossil fuels and lower emissions from biomass, renewables, or waste heat.

Regardless of which type of energy system is used, there is the possibility of integrating solar energy into the circuit. The most widespread solution is for the production of solar thermal energy to be consumed in the building itself, without exporting it to the grid. How it is adopted depends on the configuration of the overall system, where the working temperatures of the heating network play a very important role.

When it comes to cooling, electric power compression systems are most commonly used. There is the possibility of using absorption and adsorption systems that are powered by heat sources. It is suitable to integrate this technology in systems where heat generation comes from a residual source such as incinerators, waste heat, or even cogeneration. In any case, these systems need to be supported by compression cooling systems. Figure 4 shows a model of a district heating system.

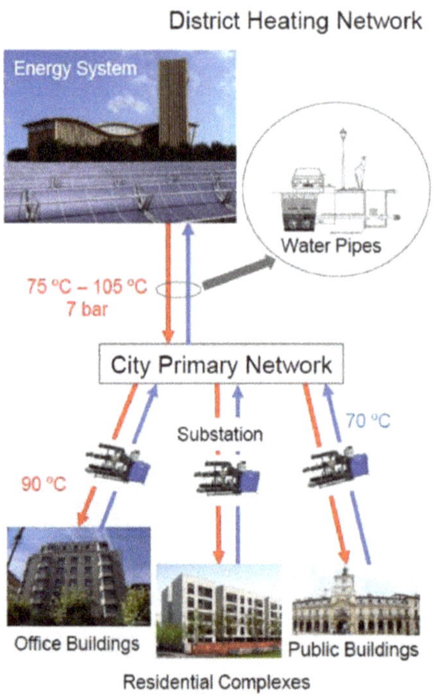

Figure 4. District heating system. Adapted from: [59].

As far as the distribution network is concerned, the ducts of DHC systems are composed of two pipelines, one for supply and one for return. The size of the system and the number of branches it has depends on its location in the energy system, the number and distribution of users, and the loss of energy to the grid. Inadequate distribution of the network can jeopardize the project's economic profitability.

Insulation is a major feature of the pipes as it is necessary to reduce to a minimum any heat loss through distribution. Pre-insulated pipes are typically used in order to avoid any problems caused by defective installation of the insulation. Cooling pipes require larger diameters due to the lower thermal gap.

There are several ways to regulate the flow rate in a pipe network; choosing one system or another depends on factors such as the type of flow rates to work with, investment costs, efficiency, operating speed, and maintenance, among others. These flow regulation systems can be either valves or multispeed pumps. The latter is the most expensive method to implement, but also the most energy-efficient and economical.

Finally, the connection to the customers and the substation consists of linking the energy distribution system, i.e., the network, to the consumers (buildings or other facilities). Water supply lines are the connection pipes between the network and the customer's substation, usually running into the customer's building from below street level. Substations adapt the distribution network pressure and temperature to the requirements of the building, guaranteeing the necessary temperature jumps for proper system efficiency.

Substations comprise equipment for regulation and control, counting equipment, and, depending on the type of substation, exchange or storage equipment. All connections to customers must be equipped with thermal energy meters. Overall, output levels of DH and industrial thermal energy production systems continue to increase, which allows for increasingly larger-scale benefits. Nevertheless, the creation of a common European framework for legislation regarding DH/DHC networks and technologies remains a vitally important objective in the EU.

3.2.3. System Integration

Secondary or residual heat often originates from industrial processes, agricultural production, and/or waste combustion ("Waste to Energy") and can be obtained either directly from the source or in conjunction with electricity through CHP systems.

One of the biggest advantages of CHP systems is the production of electricity. Different works have studied the possibility of implementing these types of systems in Spain. Paredes-Sánchez et al. [2] demonstrated the complexity involved in the development of nodal systems, which have proved their capability to adequately supply both heating and electricity through the use of conventional organic Rankine cycles (ORC). Moreover, there are also examples of heating networks using residual heat from conventional Rankine cycles currently operating in thermal power plants. Rodríguez et al. [60] coined the term "city water heating", which refers to transferring an amount of residual thermal energy (residual heat) originating in an electrical power plant to a nearby city, thus heating the drinking water supply.

The three main advantages of this type of system include overall energy savings and cost reductions for residents, zero contribution to global warming, and a significant reduction in CO_2 emissions. Moreover, many state-of-the-art CHP systems are also capable of recovering residual heat from the system to provide cooling, heating, and energy [61,62].

Heating from renewable sources without the combustion process is an option in energy systems. Solar heating can be readily adapted to both small- and large-scale systems. Currently, large solar energy farms are an important part of urban heating systems in Denmark [63], even though the conditions for solar radiation energy are generally not as favourable in higher latitudes compared to other regions like Spain.

The advantages of thermal systems using renewable energy sources include significant reductions in both fossil fuel consumption and GHG emissions by helping facilitate the transition to a highly efficient and renewable energy source in the future [2].

Particular mention should be made of the use of heat pumps in these systems. Heat pumps are generally electrically operated during periods of surplus electrical generation, for example, in Scandinavia when there is a surplus of wind energy. COP is defined as the proportion of heat supplied to the DHC system and the electricity that is consumed.

The use of heat pumps in urban heating systems is one of the most promising advances for improving overall energy efficiency, and the economic figures will play a key role in meeting the current European energy and climate objectives established for 2030 and 2050. In order to determine the correct positioning, connection, and operating modes of heat pumps, it is essential to evaluate the available data and seek out the experience and training of both city planners and engineers, as these systems must satisfy the demand for heat while operational [64]. The COP of heat pumps oscillates between 2.5 and 5.5, depending on factors including the cooling levels and temperature of the lower source, the characteristics of the carrier fluid being used, and the temperature range of the higher source. The COP of absorption heat pumps oscillates between 1.7 and 2.3 in two-stage systems, which require steam, gas, or high-temperature water as a lower energy heat source [7]. De Carli et al. [65] demonstrated that heat pumps, with or without the support of solar panels, can reduce primary energy consumption by 50–60% compared to standard systems, and a combination of heat pumps and boilers can reduce it by an additional 30–35%, which highlights the importance of adopting hybrid energy technologies in the future.

Conventional boilers are often used as a backup whenever an excess of energy is produced. Many different types of fuels can be used in these systems (including biomass) with a thermal efficiency ranging from 0.85–0.97 [7]. However, even higher efficiency levels are possible when gas-fired boilers are used in conjunction with exhaust condensation techniques. The use of burning combustibles for heat production has been widely studied. Paredes-Sánchez et al. [24] analysed the importance of biomass use in heating networks by defining District Bioheating Systems (DBS), which underscored the importance of utilising previously unused energy sources to reduce CO_2 emissions. The above indicated that the criteria most frequently used in the classification process include the morphology of the system, the services offered, and the profile of the clients However, given that micronetworks involve smaller-scale geographical areas (limited network extensions) and have fewer clients, the classification criterion of services offered was used for this study.

In addition to the technology used in obtaining thermal energy, the so-called Industry 4.0 has ushered in a wealth of benefits for the production and energy sectors alike. Within the production industry, in particular, elements such as device identification, cloud connectivity, and AI device support systems have offered substantial benefits to both the overall service and the end users by making significant improvements to energy efficiency, final energy cost, and quality of the energy supply. Thermal operation optimisation, which is a process using artificial intelligence technology to perform a specific task with a specific objective, plays an important role in finding the optimal balance between the energy temperature and flow within a district system to minimize costs and ensure the quality of the energy supply.

3.3. *Phase III: Energy Management*

Based on the previous analysis of energy systems (i.e., Sections 3.2.1–3.2.3), one of the main benefits of DHC systems is their ability to offer higher-efficiency energy production by integrating a variety of renewable energy sources (biomass, geothermal, thermal solar, etc.) and local resources that would otherwise go unused (natural cooling, excess heat or cooling from nearby industrial work, integration of both heating and cooling, etc.).

A combination of the aforementioned factors, along with appropriate energy management of the generation/demand binominal and continual professional maintenance and management, can significantly contribute to reducing energy consumption levels, CO_2 emissions, and air pollution while providing a highly stable energy supply.

There are many important parties involved in the successful completion of DHC projects, including local administrations, installation management companies (generally energy service providers), energy company industries, suppliers, property developers, and end clients. The respective city halls and city councillors play especially influential roles, as they are responsible for territorial planning. Moreover, administrations can further facilitate the administrative process by either approving or rejecting projects, making economic contributions, and taking a more active role in the process.

The ability to identify an opportunity to develop a district heating system in the urban planning stage is key to the success of the project, as it helps to reduce costs and allows for easier integration of other services. System costs include updating both the existing heating systems and the heating distribution networks, which in turn minimises energy loss within the system, promotes more efficient use of low-temperature energy sources and higher overall efficiency, and, most importantly, allows for greater integration of other systems when compared to DH systems [66]. Thus, the next step in the development of these systems will be to conduct a study of their economic viability to provide a profit analysis to help in the final decision-making process.

Despite the possible technological impediments, legislative issues, or network management difficulties, the expertise offered by the ongoing work of experienced companies in the sector engaged in fully operational networks makes the logistics of such project ever less complicated.

The current management model for DHC projects in Spain relies on joint ventures between both private and public entities. Likewise, public agencies, associations, and institutions responsible for promoting and developing energy-efficient technologies also play an important role in helping secure resources like subsidies as well as promoting the use of DHC systems in municipal and regional energy plans.

The ability to secure financing and installation management services is also a crucial factor in determining the viability of a project. The joint venture is responsible for securing the necessary financial backing for the project. Energy service companies manage the facilities themselves while providing know-how in the construction process and subsequent management of any resources involved in the commercialisation and operation of the facilities. Lastly, the end client is also a determining factor in the successful achievement of a project.

When dealing with a new urban project, the connection timeline for prospective clients is a key issue, and the planning and design previsions must be as realistic as possible. In those cases where previously inhabited urban areas are involved, the local authorities play an important role in the planning process as they have the final decision regarding the approval and execution of the project. Regardless of the particulars of any given individual project, however, strict adherence to the life cycle for the installations is essential.

Based on methodological analysis [31,33], Spain has an estimated total output of 1448 MW_{th}, 72% of which is supplied exclusively with renewable energy sources. The remaining 28% come from a combination of energy sources, with natural gas being the most common in Spain. The total registered output includes 612 MW_{th} from heating networks, 829 MW_{th} from heating and cooling systems, and 7 MW_{th} from cooling (Figure 5).

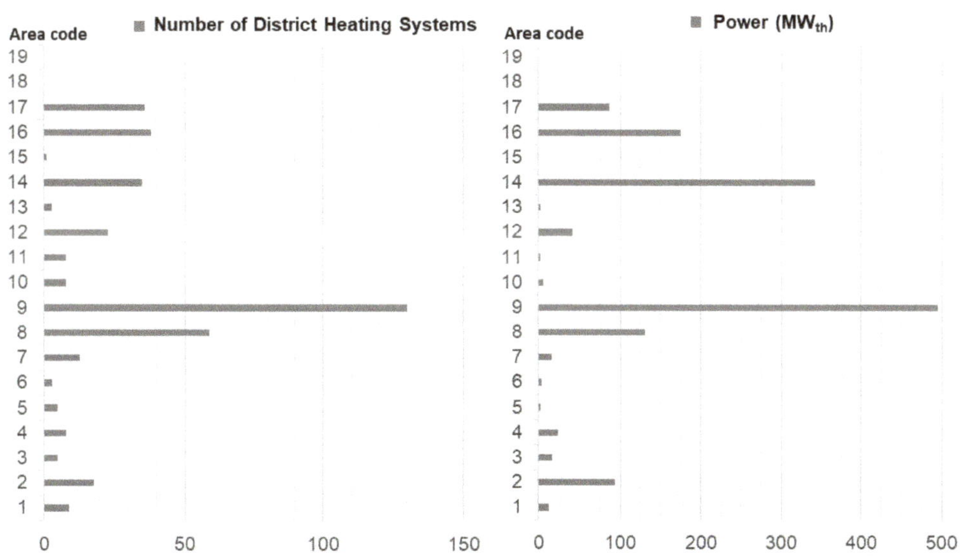

Figure 5. Number of DH/DHC systems and thermal power output by area code.

These systems offer some decarbonisation measures, including their abilities to use more efficient technologies, replace coal with lower-contamination fossil fuel sources like natural gas, and run exclusively on renewable energy sources [67]. Zones with area codes 2, 3, 7, and 8 (Table 2) are heavily dependent on coal industry in thermal applications, which could make the development of DH/DHC based on biomass an important energy goal.

Thus, biomass and natural gas (or a combination of the two) provide 63% of total output, with 20% coming directly from renewable sources. Specifically, biomass is used, either exclusively or in combination with other fuel sources, in 3 out of every 4 networks. In terms of total energy output, 73% is used for heating and 27% for cooling.

Overall, district energy output has been continually increasing in recent years [31]. Catalonia (495 MW_{th}), Community of Madrid (342 MW_{th}), and Autonomous Community of Navarre (175 MW_{th}), area codes 9, 14, and 16 in Table 2, respectively, contribute approximately 70% of the national output in Spain. In terms of network type, 363 provide heating, 35 provide heating and cooling, and 4 provide cooling (Figures 6 and 7).

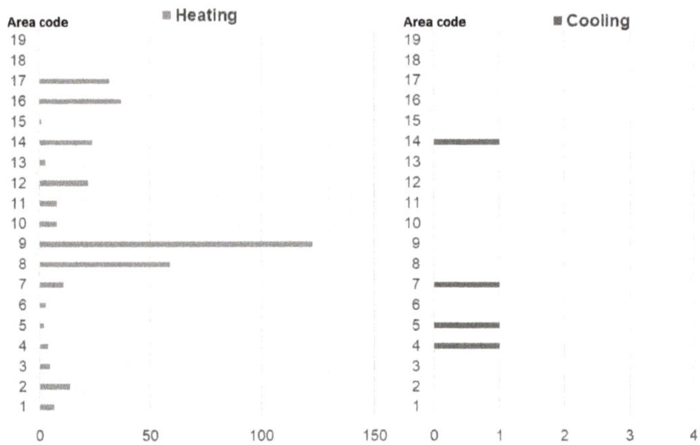

Figure 6. Number of district heating or cooling systems by area code.

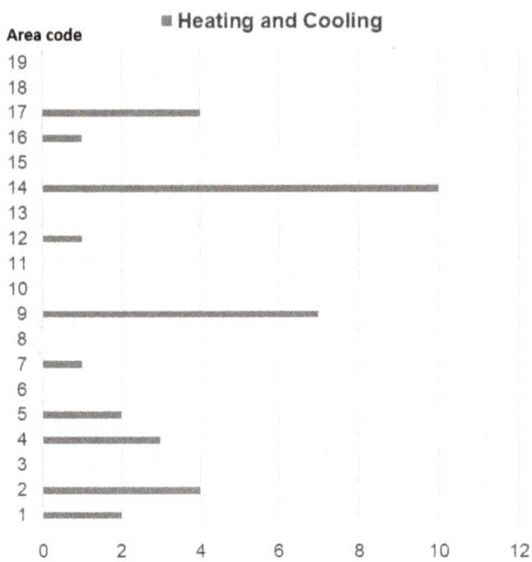

Figure 7. Number of district heating and cooling systems by area code.

In terms of client profile, 68% belong to the services sector, 24% to housing, and 8% to industry. In terms of total consumption, 45% comes from the services sector, 32% from housing, and 23% from industry.

According to the data, 49% of Spanish networks are public property, 47% are private, and 4% are mixed. By region, Catalonia registers the highest number of DHC networks (130), followed by Castile and Leon (59), the Autonomous Community of Navarre (38), La Rioja, and Cantabria (3 each), and the Region of Murcia (1), area codes 9, 8, 16, 13, and 15 in Table 2.

District heating systems in Spain have an overall thermal efficiency in boilers of around 90% and COP of 4 in heat pumps. The total thermal production corresponds to 1448 MW_{th}, of which 72% is exclusively supplied by renewable energy sources. The Association of District Heating and Cooling Companies [31] has registered 402 networks in Spain, servicing a total of 5000 buildings with an estimated network of approximately 680 km. Combined, these networks account for an annual reduction of 305,945 tonnes of CO_2 emissions and a 79% reduction in fossil fuel use. The existing DHC in each zone are studied based on the operational information available in the databases, as described in Section 2. In this respect, homogenisation of this information is pursued by means of a particular analysis of the behaviour and barriers included in Table 3. One of the most important DHC networks in Spain is located in Barcelona (area code 9, Table 2), which was built in 2002 in order to provide heating, air conditioning, and DHW. It extends over a distance of nearly 18 km and provides services to 100 buildings [68].

In addition, the City Hall of Barcelona, together with the Public Consortium Local Energy Agency of Barcelona, planned to develop an 18,000 home residential complex near the Seat Automotive plant in the Zona Franca district of Barcelona (Hospitalet City Hall). However, due to the global economic crisis, the project was never realised. What made this particular project noteworthy was the fact that its trigeneration power plant was to be equipped with heating and cooling systems, a glycol cooling bath (−10 °C), water source heat pumps, and photovoltaic solar panels, and it was configured to service over 1,200,000 m^2 [69].

The Mostoles district in Madrid (Community of Madrid) (area code 14, Table 2) offers an example of an exclusively urban area heating and DHW network using biomass (pellets and pruning waste). A total of 3000 homes have already been connected to the heating network during the initial stage of the project, and the total will increase to 6000 homes in subsequent stages [70].

Industrial activities (e.g., technology centres, industrial complexes, etc.) are often located in or near mining areas and can help minimize the ratio between distance and consumption, which is essential in these projects [33]. The primary technical challenge involves the transportation of thermal energy over large distances, since end consumers are often located a considerable distance from production centres, and very few mining areas are close to urban areas.

Area code 3 offers noteworthy examples of the progress being made in geothermal technology, including a project developed by Grupo Hunosa, which uses mining water for a DH system located in Mieres (Principality of Asturias) (area code 3, Table 2). The network originates at the mining area of the Pozo Barredo and provides service to the Polytechnic School of Mieres of the University of Oviedo, Bernaldo de Quirós High School, and a group of buildings containing 248 homes located in the Vasco-Mayacina neighbourhood [71]. Public institutions must play a pivotal role in initial contract negotiations to encourage private enterprises to participate in these types of network projects in the future. In this respect, it is worth mentioning that district heating has been installed, with similar circumstances, in the Pontevedra campus of the University of Vigo. This infrastructure connects the Faculty of Education and Sport Sciences with the Faculty of Social Sciences and Communication. The thermal system has two combustion chambers of movable grate of 1 MW_{th}, each of which is supplied with wood chips (Figure 8).

Figure 8. Thermal system in the Pontevedra campus of the University of Vigo.

In addition, at the Campus of Ourense, a project of geothermal installation to meet the demand for heating (80%) and cooling (100%) by a "hybrid" system stands out. This district heating is a combination of aerothermal production of about 200 kWth and geothermal generation of approximately 500 kWth. The thermal system has five heat pumps in cascade configuration.

As far as research into heating networks is concerned, a lot of progress has been made recently in Spain. Fourth-generation urban heating systems (4GDH) are being discussed now, an example of which is the SmartEnCity project sponsored by Vitoria-Gasteiz (Basque Country) (area code 17, Table 2) City Hall as part of the EU Horizon 2020 program. The idea behind the project is that if more citizens become actively involved in the planning process, fewer people will reject the idea of heating networks in the future.

The SmartEnCity project calls for the complete renovation of the entire Coronación neighbourhood and the creation of a biomass-based network that is capable of meeting the basic energy needs of 750 to 1313 homes. Additionally, it will develop integrated thermal and electric infrastructures, encourage sustainable mobility by using cleaner technologies in vehicle fleets, help spread technologies of the information and communication, and promote urban renewal by renovating public spaces like streets and squares [72].

In addition, the R2CITIES project was born in Spain and at the international level, the objective of which is to create and develop repeatable, large-scale renovation projects for the construction and management of district heating to achieve cities with near-zero energy consumption. Currently, pilot programs are operating in Kartal (Turkey), Valladolid (Spain), and Genoa (Italy), all of which have different climates and objectives. Each of the three programmes is being managed by its respective municipality; these municipalities are also the principal promoters of these highly ambitious neighbourhood renovation programmes. Single projects prove that a systematic approach, in combination with the use of technologies such as insulation and information and communication technologies, as well as cost-effective and energy-efficient resources, can not only achieve excellent results in terms of energy efficiency but also drastically reduce CO_2 emissions [73].

The DHC engineering firm DH Eco Energías has initiated a "macro-project" in Spain for the promotion and construction of hybrid networks with biomass and concentrated solar heating systems on a budget of EUR 204 million. Thanks to the positive environmental impact of the project, the Ministry of Ecological Transition has selected it to become part of the "Climate Project", since it will prevent 360,000 t of CO_2 from being released annually, which is equal to the pollution from some 240,000 vehicles. The ten heating network systems involved in the project will generate 1335 jobs in total during the construction phase and will be developed in 10 different locations throughout Spain, including Ávila, Huesca, Oviedo, Palencia, Salamanca, Valladolid, Zamora, Boadilla del Monte, Coslada, and Leganés. It could provide service to a total of 111,545 homes and cover an annual energy demand of 1100 GWh [59].

4. Discussion

The institutional context takes into account factors including the basic drive forces from resources to energy management, the importance of awareness of the economic benefits of district heating and cooling systems, ownership, legal frameworks, prices, and advancement in knowledge [10]. The majority of the barriers currently facing DHC systems arise during the initial planning and proposal stages. These barriers come from different fields and are often of technical, economic, institutional, social or cultural, institutional, and legal natures.

Technical barriers mainly arise during construction of, e.g., the energy system or building structure. District heating networks involve some factors such as:

- The installation of a heating production system using existing technology.
- The needs of large-scale civil engineering projects, which vary greatly according to the scope of the project.

The latter factor is accentuated when providing services in populated urban areas, where street work is required and often disrupts other services.

According to the technical building codes responsible for the certification process in buildings, there are no standards for rewarding buildings that are serviced by district networks. However, there are currently proposals aimed at remedying this unfortunate situation. At present, all new buildings must satisfy a portion of their hot water energy demand through either thermal solar energy or an equally efficient, previously approved alternative.

This regulation causes a challenge with district network systems that do not incorporate CHP, residual heat, or alternative systems providing equivalent energy savings. At present, there is no legal recourse to remedy this situation, even though logic would dictate that buildings serviced by district heating networks should be exempt from such legislation.

However, if the district network is not supplied by a renewable or residual energy source, the environmental impact could be greater. One possible solution to this problem would be changing current legislation regarding thermal solar installations, urban municipal schemes, and energy plans to include DH/DHC systems as an option in new housing projects with favourable conditions on the basis that they are a profitable and effective means of reducing energy consumption in high energy demand areas. Financial incentives should also be considered to encourage the construction of these types of systems and network connections in addition to searching for ways to ease regulations in the future.

With regard to economic barriers, they arise as a result of the size and scope of the proposed project, as the majority of them involve civil works projects that affect distribution networks. Akhtari et al. [4] outlined the numerous social and environmental advantages inherent in the use of renewable energy sources in these types of systems. However, implementing these systems in inhabited areas can exponentially increase construction costs. Moreover, recovering the initial investment costs of these types of project takes a long time for private capital funds, which means that public aid or participation from public organisations is needed to cover the initial investment and the necessary maintenance costs until the initial investment cost can be recovered.

Uncertainties in the timeframe for new client connections can complicate the task of calculating the medium-term revenues of the operating company. The timeline can also be affected by economic cycles, which means that the design process must consider a number of economic factors. In this sense, the selection of a system that provides enormous environmental benefits can ultimately become economically unfeasible, and a less expensive system can either provide little to no positive environmental impact or be environmentally harmful [29].

As previously mentioned, social or cultural barriers generally arise in projects planned for inhabited urban areas that are intended to substitute existing heating infrastructures with DH/DHC systems. Widespread unfamiliarity with the operation and management of these systems often complicates the decision to replace an existing operational heating system that is already familiar to the consumer.

The Spanish State Department and a number of regional energy agencies have recently undertaken the task of providing local authorities with informative material, including a municipal ordinance guide explaining the current legislative framework and legal guidelines for local administrations, promoters, and building developers [32].

Generally, there is less of this type of resistance in regions where network heating systems are already prevalent. DH systems are widely considered to be the simplest and most efficient way of remedying the problem of low energy efficiency in urban areas.

The key to the success of DH networks is finding the right balance between national governmental policies and local city council initiatives; they are the institutional barriers. Social participation, especially in the initial stages of development, is another essential part of the process. Therefore, coordination between the public and private enterprises responsible for financing, maintaining, and developing these projects is an essential part of successful policymaking.

Private funding for low carbon technologies by residents is another viable option for the construction of DH systems in open markets [74]. While governmental participation in the form of payment plans requires soft loans and other financial incentives, it reduces the financial risks involved for private investors and helps stimulate public interest in DH systems.

In many European states, local councils have more administrative control and greater financial clout [26]. Many of these public city administrations (institutional barrier) determine their energy policies to benefit the local area rather than simply seeking financial rewards, which helps ensure that these projects benefit all of society rather than private enterprises alone [75]. Moreover, this strategy helps encourage the use of local labour forces and promotes greater levels of local technical expertise.

More widespread use of these systems at a variety of levels is essential to overcome the abovementioned social barriers.

This study aims to help in this process of promoting greater public awareness (social barrier). Prospective clients need detailed explanations and/or demonstrations of these systems in order to familiarise themselves with their use and fully comprehend the financial benefits and reliability of the services they provide (social barrier).

In the social context, technicians, town planners, engineers, and public entities in charge of energy management must strive to create a better public understanding of these systems, which starts by ensuring a bigger presence in school and university curriculums. Ultimately, however, more widespread adoption of DH/DHC network systems is the first step towards creating a better general awareness of these benefits of these systems to make them more accessible to potentially interested parties.

To resolve the challenge regarding legislation (legal barrier) requiring thermal solar energy systems in new buildings and ensure flexibility, systems must possess inertia to maintain a balanced energy supply at all times. Thermal networks can use thermal energy as a source of thermal inertia. These capacities are located in different places throughout the network, including the heat/cooling carrier fluid, thermal storage reservoirs, and the thermal inertia of the buildings being serviced with heating/cooling [27].

Special attention must be given to managing thermal energy systems to reduce their carbon footprint and GHG emissions. One of the principal advantages of district heating systems is their ability to significantly reduce CO_2 emissions through the use of polygeneration energy conversion technology.

Likewise, there are clear benefits to using excess industrial heat as an energy source as it is free and can be easily integrated into urban heating systems. Moreover, there are social benefits including the reduction of pollution [76]. An analysis of the results obtained from continued improvements to urban heating systems, along with the corresponding reduction in energy demand, demonstrates that it is crucial to continue their current line of development [29]. Future challenges lie in the parametric modelling and optimisation of the individual systems, which must be developed through the analysis of case studies [77]. Additionally, this characterisation will enable the results of Phase III to be implemented parametrically as an in situ thermal system [78].

5. Conclusions

Globally, DH/DHC have very strong technical and economic potentials and represent a future viable heating and cooling supply option. However, further efforts are required to identify, assess, and implement these potentials with a view to fully harvesting the global benefits of district heating and cooling. Based on the data obtained from the analysis, there is now a good understanding of how to deal with the technical aspects of resources, technology and management for the implementation of heat networks. The present study led to the following conclusions for the implementation of heat networks:

1. Heating networks require a centralised heating source for several interconnected buildings in a given area (e.g., hospitals). What DH systems have in common is the use of a centralised heating source, which allows for the use of more efficient technologies and requires energy management services.
2. The key performance indicator for all of the technologies and energy sources discussed in the present work is the ability to successfully combine resources, technology, and energy management to available energy sources on the market.
3. Through the use of renewable energy sources alone, it is possible to reduce the amount of fossil fuel consumption. Moreover, the resulting energy savings create energy efficiency opportunities and reduce area CO_2 levels. These networks account for an annual reduction of 305,945 t of CO_2 emissions and a 79% reduction in fossil fuel use in Spain.
4. DH systems are of particular interest to European regions that are undergoing a process of energy transition. At present, zones with area codes 2, 3, 7, and 8 (Table 2; Figures 5–7) in Spain are heavily dependent on coal industry, which makes the development of district heating networks based on biomass an important technological and energetic goal.

District heating and cooling networks combine a wide variety of technological solutions and energy management strategies. Properly organized district heating networks ultimately provide many benefits to all of the parties involved, including public administrations, energy service providers, property developers, and end-users, among others. The supplied energy needs to meet both quality and energy efficiency standards while remaining economically viable. Therefore, the methodology presented in this study provides a very powerful decision-making tool for thermal energy systems. The main challenge now is understanding the specific local parameters, operational conditions, and legal framework.

Author Contributions: Conceptualization and methodology, B.M.P.-S. and J.P.P.; writing—original draft preparation, B.M.P.-S.; writing—review and editing, B.M.P.-S.; visualization, J.P.P., N.C., and E.R.-L.; supervision, B.M.P.-S., J.P.P., N.C., and E.R.-L. All authors have read and agreed to the published version of the manuscript.

Funding: The authors would like to express their acknowledgment to the Biomasa-AP project, which was approved by the INTERREG V-A España–Portugal 2014–2020 programme (POCTEP) and cofinanced by the European Regional Development Fund (FEDER) and the H2020-FTI-Pilot-2016-1-760551-CYCLOMB approved by the European Commission. We would like to thank all the companies and institutions that have contributed in some way to the work described in this paper, and especially DH Eco Energías (FUO-19-118) for its research collaboration with the University of Oviedo.

Institutional Review Board Statement: Not applicable.

Informed Consent Statement: Not applicable.

Data Availability Statement: Not applicable.

Conflicts of Interest: The authors declare no conflict of interest.

References

1. UNDP. Human Development Report 2019. Available online: http://hdr.undp.org/sites/default/files/hdr2019.pdf (accessed on 7 April 2021).
2. Paredes-Sánchez, J.P.; López-Ochoa, L.M. Bioenergy as an Alternative to Fossil Fuels in Thermal Systems. In *Advances in Sustainable Energy. Lecture Notes in Energy*; Vasel-Be-Hagh, A., Ting, D., Eds.; Springer: Cham, Switzerland, 2019; Volume 70, pp. 149–168.
3. Míguez, J.L.; Porteiro, J.; Pérez-Orozco, R.; Patiño, D.; Rodríguez, S. Evolution of CO_2 capture technology between 2007 and 2017 through the study of patent activity. *Appl. Energy* **2018**, *211*, 1282–1296. [CrossRef]
4. Akhtari, S.; Sowlati, T.; Day, K. Economic feasibility of utilizing forest biomass in district energy systems—A review. *Renew. Sustain. Energy Rev.* **2014**, *33*, 117–127. [CrossRef]
5. INE. Database Report 2019. [In Spanish]. Available online: https://www.ine.es/en/index.htm (accessed on 7 April 2021).
6. Jana, K.; Ray, A.; Majoumerd, M.M.; Assadi, M.; De, S. Polygeneration as a future sustainable energy solution—A comprehensive review. *Appl. Energy* **2017**, *202*, 88–111. [CrossRef]
7. Sayegh, M.A.; Danielewicz, J.; Nannou, T.; Miniewicz, M.; Jadwiszczak, P.; Piekarska, K.; Jouhara, H. Trends of European research and development in district heating technologies. *Renew. Sustain. Energy Rev.* **2017**, *68*, 1183–1192. [CrossRef]
8. EurObserv'ER. Database. Available online: https://www.eurobserv-er.org/pdf/18th-annual-overview-barometer/ (accessed on 7 April 2021).
9. Benedek, J.; Sebestyén, T.T.; Bartók, B. Evaluation of renewable energy sources in peripheral areas and renewable energy-based rural development. *Renew. Sustain. Energy Rev.* **2018**, *90*, 516–535. [CrossRef]
10. Werner, S. International review of district heating and cooling. *Energy* **2017**, *137*, 617–631. [CrossRef]
11. Buffa, S.; Cozzini, M.; D'Antoni, M.; Baratieri, M.; Fedrizzi, R. 5th generation district heating and cooling systems: A review of existing cases in Europe. *Renew. Sustain. Energy Rev.* **2019**, *104*, 504–522. [CrossRef]
12. Rezaie, B.; Rosen, M.A. District heating and cooling: Review of technology and potential enhancements. *Appl. Energy* **2012**, *93*, 2–10. [CrossRef]
13. Levihn, F. CHP and heat pumps to balance renewable power production: Lessons from the district heating network in Stockholm. *Energy* **2017**, *137*, 670–678. [CrossRef]
14. Averfalk, H.; Ingvarsson, P.; Persson, U.; Gong, M.; Werner, S. Large heat pumps in Swedish district heating systems. *Renew. Sustain. Energy Rev.* **2017**, *79*, 1275–1284. [CrossRef]
15. Sarbu, I.; Sebarchievici, C. A Comprehensive Review of Thermal Energy Storage. *Sustainability* **2018**, *10*, 191. [CrossRef]
16. Dodds, P.E.; Staffell, I.; Hawkes, A.D.; Li, F.; Grünewald, P.; McDowall, W.; Ekins, P. Hydrogen and fuel cell technologies for heating: A review. *Int. J. Hydrogen Energy* **2015**, *40*, 2065–2083. [CrossRef]
17. Fragiacomo, P.; De Lorenzo, G.; Corigliano, O. Performance analysis of a Solid Oxide Fuel Cell-Gasifier integrated system in co-trigenerative arrangement. *J. Energy Resour. Technol.* **2018**, *140*, 1–9. [CrossRef]
18. Palomba, V.; Prestipino, M.; Galvagno, A. Tri-generation for industrial applications: Development of a simulation model for a gasification-SOFC based system. *Int. J. Hydrog. Energy* **2017**, *42*, 27866–27883. [CrossRef]
19. APPA. Study of the Macroeconomic Impact of Renewable Energies in Spain. [In Spanish]. Available online: https://www.appa.es/wp-content/uploads/2018/10/Estudio_del_impacto_Macroeconomico_de_las_energias_renovables_en_Espa%C3%B1a_2017.pdf (accessed on 7 April 2021).
20. BIOPLAT. The Bioenergy Sector in Spain. Available online: http://bioplat.org/download/7721 (accessed on 7 April 2021).
21. Silva, B.N.; Khan, M.; Han, K. Towards sustainable smart cities: A review of trends, architectures, components, and open challenges in smart cities. *Sustain. Cities Soc.* **2018**, *38*, 679–713. [CrossRef]
22. Ericsson, K.; Werner, S. The introduction and expansion of biomass use in Swedish district heating systems. *Biomass Bioenergy* **2016**, *94*, 57–65. [CrossRef]
23. Pinto, J.F.; da Graça, G.C. Comparison between geothermal district heating and deep energy refurbishment of residential building districts. *Sustain. Cities Soc.* **2018**, *38*, 309–324. [CrossRef]
24. Paredes-Sánchez, J.P.; López-Ochoa, L.M.; López-González, L.M.; Xiberta-Bernat, J. Bioenergy for District Bioheating System (DBS) from eucalyptus residues in a European coal-producing region. *Energy Convers. Manag.* **2016**, *126*, 960–970. [CrossRef]
25. Paredes-Sánchez, J.P.; Conde, M.; Gómez, M.A.; Alves, D. Modelling hybrid thermal systems for district heating: A pilot project in wood transformation industry. *J. Clean. Prod.* **2018**, *194*, 726–734. [CrossRef]

26. Mazhar, A.R.; Liu, S.; Shukla, A. A state of art review on the district heating systems. *Renew. Sustain. Energy Rev.* **2018**, *96*, 420–439. [CrossRef]
27. Vandermeulen, A.; van Der Heijde, B.; Helsen, L. Controlling district heating and cooling networks to unlock flexibility: A review. *Energy* **2018**, *151*, 103–115. [CrossRef]
28. Lund, H.; Duic, N.; Alberg, P.; Vad, B. Future district heating systems and technologies: On the role of smart energy systems and 4th generation district heating. *Energy* **2018**, *165*, 614–619. [CrossRef]
29. Lake, A.; Rezaie, B.; Beyerlein, S. Review of district heating and cooling systems for a sustainable future. *Renew. Sustain. Energy Rev.* **2017**, *67*, 417–425. [CrossRef]
30. Paredes-Sánchez, J.P.; López-Ochoa, L.M.; López-González, L.M.; Las-Heras-Casas, J.; Xiberta-Bernat, J. Evolution and perspectives of the bioenergy applications in Spain. *J. Clean. Prod.* **2019**, *213*, 553–568. [CrossRef]
31. ADHAC. Census of Heat and Cold Networks 2018. [In Spanish]. Available online: www.adhac.es/Priv/ClientsImages/AsociacionPerso8_1540226118.pdf (accessed on 7 April 2021).
32. IDAE. Renewable Energy Plan (PER). [In Spanish]. Available online: https://www.idae.es/tecnologias/energias-renovables/plan-de-energias-renovables-2011-2020 (accessed on 7 April 2021).
33. RHC. 100% Renewable Energy Districts: 2050 Vision. [In Spanish]. Available online: https://www.euroheat.org/wp-content/uploads/2019/08/RHC-ETIP_District-and-DHC-Vision-2050.pdf (accessed on 7 April 2021).
34. Cacabelos, A.; Eguía, P.; Míguez, J.L.; Granada, E.; Arce, M.E. Calibrated simulation of a public library HVAC system with a ground-source heat pump and a radiant floor using TRNSYS and GenOpt. *Energy Build.* **2015**, *108*, 114–126. [CrossRef]
35. Kerimray, A.; Rojas-Solórzano, L.; Torkmahalleh, M.A.; Hopke, P.K.; Gallachóir, B.P.Ó. Coal use for residential heating: Patterns, health implications and lessons learned. *Energy Sustain. Dev.* **2017**, *40*, 19–30. [CrossRef]
36. Paredes-Sánchez, B.M.; Paredes-Sánchez, J.P.; García-Nieto, P.J. Energy Multiphase Model for Biocoal Conversion Systems by Means of a Nodal Network. *Energies* **2020**, *13*, 2728. [CrossRef]
37. Ghisellini, P.; Cialani, C.; Ulgiati, S. A review on circular economy: The expected transition to a balanced interplay of environmental and economic systems. *J. Clean. Prod.* **2016**, *114*, 11–32. [CrossRef]
38. Hagos, D.A.; Gebremedhin, A.; Bolkesjø, T.F. The prospects of bioenergy in the future energy system of Inland Norway. *Energy* **2017**, *121*, 78–91. [CrossRef]
39. Lausselet, C.; Cherubini, F.; David, G.; Serrano, A.; Becidan, M.; Hu, X.; Rørstad, P.K.; Strømman, A.H. Norwegian Waste-to-Energy: Climate change, circular economy and carbon capture and storage. *Resour. Conserv. Recycl.* **2017**, *126*, 50–61. [CrossRef]
40. Nieto, P.G.; García–Gonzalo, E.; Lasheras, F.S.; Paredes–Sánchez, J.P.; Fernández, P.R. Forecast of the higher heating value in biomass torrefaction by means of machine learning techniques. *J. Comput. Appl. Math.* **2019**, *357*, 284–301. [CrossRef]
41. Paredes-Sánchez, J.P.; Gutiérrez-Trashorras, A.J.; Xiberta-Bernat, J. Wood residue to energy from forests in the Central Metropolitan Area of Asturias (NW Spain). *Urban For. Urban Green.* **2015**, *14*, 195–199. [CrossRef]
42. Paredes-Sánchez, J.P.; Gutiérrez-Trashorras, A.J.; Xiberta-Bernat, J. Energy potential of residue from wood transformation industry in the central metropolitan area of the Principality of Asturias (northwest Spain). *Waste Manag. Res.* **2014**, *32*, 241–244. [CrossRef] [PubMed]
43. Las-Heras-Casas, J.; López-Ochoa, L.M.; López-González, L.M.; Paredes-Sánchez, J.P. A tool for verifying energy performance certificates and improving the knowledge of the residential sector: A case study of the Autonomous Community of Aragón (Spain). *Sustain. Cities Soc.* **2018**, *41*, 62–72. [CrossRef]
44. Fernández, S.T.; Sánchez, J.P.P.; Trashorras, A.J.G. Analysis of forest residual biomass potential for bioenergy production in Spain. *Clean Technol. Environ. Policy* **2016**, *18*, 209–218. [CrossRef]
45. Huang, J.; Fan, J.; Furbo, S. Feasibility study on solar district heating in China. *Renew. Sustain. Energy Rev.* **2019**, *108*, 53–64. [CrossRef]
46. Hansen, K.; Mathiesen, B.V. Comprehensive assessment of the role and potential for solar thermal in future energy systems. *Solar Energy* **2018**, *169*, 144–152. [CrossRef]
47. Paredes-Sánchez, J.P.; Las-Heras-Casas, J.; Paredes-Sánchez, B.M. Solar Energy, the Future Ahead. In *Advances in Sustainable Energy. Lecture Notes in Energy*; Vasel-Be-Hagh, A., Ting, D., Eds.; Springer: Cham, Switzerland, 2019; Volume 70, pp. 149–168.
48. Huculak, M.; Jarczewski, W.; Dej, M. Economic aspects of the use of deep geothermal heat in district heating in Poland. *Renew. Sustain. Energy Rev.* **2015**, *49*, 29–40. [CrossRef]
49. Corigliano, O.; Fragiacomo, P. Extensive analysis of SOFC fed by direct syngas at different anodic compositions by using two numerical approaches. *Energy Convers. Manag.* **2020**, *209*, 112664. [CrossRef]
50. Thorsteinsson, H.H.; Tester, J.W. Barriers and enablers to geothermal district heating system development in the United States. *Energy Policy* **2010**, *38*, 803–813. [CrossRef]
51. Ericsson, K.; Nilsson, L.J. Assessment of the potential biomass supply in Europe using a resource-focused approach. *Biomass Bioenergy* **2006**, *30*, 1–15. [CrossRef]
52. Su, D.; Zhang, Q.; Wang, G.; Li, H. Market analysis of natural gas for district heating in China. *Energy Procedia* **2015**, *75*, 2713–2717. [CrossRef]
53. Urbaneck, T.; Oppelt, T.; Platzer, B.; Frey, H.; Uhlig, U.; Göschel, T.; Zimmermann, D.; Rabe, D. Solar district heating in East Germany–transformation in a cogeneration dominated city. *Energy Procedia* **2015**, *70*, 587–594. [CrossRef]

54. Vakkilainen, E.K. *Steam Generation from Biomass: Construction and Design of Large Boilers*; Butterworth-Heinemann: Oxford, UK, 2017; p. 302.
55. Ozgener, L.; Hepbasli, A.; Dincer, I. A key review on performance improvement aspects of geothermal district heating systems and applications. *Renew. Sustain. Energy Rev.* **2007**, *11*, 1675–1697. [CrossRef]
56. Sayigh, A. *Comprehensive Renewable Energy*; Elsevier: Amsterdam, The Netherlands, 2012; p. 4422.
57. Fazlollahi, S.; Becker, G.; Maréchal, F. Multi-objectives, multi-period optimization of district energy systems: II—Daily thermal storage. *Comput. Chem. Eng.* **2014**, *71*, 648–662. [CrossRef]
58. Parameshwaran, R.; Kalaiselvam, S.; Harikrishnan, S.; Elayaperumal, A. Sustainable thermal energy storage technologies for buildings: A review. *Renew. Sustain. Energy Rev.* **2012**, *16*, 2394–2433. [CrossRef]
59. DH Eco Energías. Technical Report. [In Spanish]. Available online: http://www.dhecoenergias.es/ (accessed on 7 April 2021).
60. Rodríguez, R.; Bello, V.G.; Díaz-Aguado, M.B. Application of eco-efficiency in a coal-burning power plant benefitting both the environment and citizens: Design of a "city water heating" system. *Appl. Energy* **2017**, *189*, 789–799. [CrossRef]
61. Liu, M.; Shi, Y.; Fang, F. Combined cooling, heating and power systems: A survey. *Renew. Sustain. Energy Rev.* **2014**, *35*, 1–22. [CrossRef]
62. Li, Y.; Fu, L.; Zhang, S. Technology application of district heating system with Co-generation based on absorption heat exchange. *Energy* **2015**, *90*, 663–670. [CrossRef]
63. Tian, Z.; Zhang, S.; Deng, J.; Fan, J.; Huang, J.; Kong, W.; Perers, B.; Furbo, S. Large-scale solar district heating plants in Danish smart thermal grid: Developments and recent trends. *Energy Convers. Manag.* **2019**, *189*, 67–80. [CrossRef]
64. Sayegh, M.A.; Jadwiszczak, P.; Axcell, B.P.; Niemierka, E.; Bryś, K.; Jouhara, H. Heat pump placement, connection and operational modes in European district heating. *Energy Build.* **2018**, *166*, 122–144. [CrossRef]
65. De Carli, M.; Galgaro, A.; Pasqualetto, M.; Zarrella, A. Energetic and economic aspects of a heating and cooling district in a mild climate based on closed loop ground source heat pump. *Appl. Therm. Eng.* **2014**, *71*, 895–904. [CrossRef]
66. Lund, H.; Østergaard, P.A.; Chang, M.; Werner, S.; Svendsen, S.; Sorknæs, P.; Thorsen, J.E.; Hvelplund, F.; Mortensen, B.O.G.; Mathiesen, B.V. The status of 4th generation district heating: Research and results. *Energy* **2018**, *164*, 147–159. [CrossRef]
67. Hast, A.; Syri, S.; Lekavičius, V.; Galinis, A. District heating in cities as a part of low-carbon energy system. *Energy* **2018**, *152*, 627–639. [CrossRef]
68. Districlima. Districlima Technical Documentation (2019). [In Spanish]. Available online: http://www.districlima.com/es/cms/nuestra-red-nuestros-clientes (accessed on 7 April 2021).
69. Econenergies. Econenergies Technical Report. [In Spanish]. Available online: http://www.ecoenergies.cat/es (accessed on 7 April 2021).
70. Mostolesdistrictheating. Mostolesdistrictheating Technical Documentation. Available online: http://www.mostolesdistrictheating.info/ (accessed on 7 April 2021).
71. Hunosa. Hunosa Technical Documentation. [In Spanish]. Available online: https://www.hunosa.es/metaspace/portal/47635/50404-district-heating-pozo-barredo?pms=1,49877,49879004,view,normal,0 (accessed on 7 April 2021).
72. SmartEncity. Technical Report. [In Spanish]. Available online: https://smartencity.eu/ (accessed on 7 April 2021).
73. R2cities. Technical Documentation. [In Spanish]. Available online: http://r2cities.eu/ (accessed on 7 April 2021).
74. Curtin, J.; McInerney, C.; Gallachóir, B.Ó. Financial incentives to mobilise local citizens as investors in low-carbon technologies: A systematic literature review. *Renew. Sustain. Energy Rev.* **2017**, *75*, 534–547. [CrossRef]
75. Lucia, U. Econophysics and bio-chemical engineering thermodynamics: The exergetic analysis of a municipality. *Physica A* **2016**, *462*, 421–430. [CrossRef]
76. Petrov, O.; Bi, X.; Lau, A. Impact assessment of biomass-based district heating systems in densely populated communities. Part I: Dynamic intake fraction methodology. *Atmos. Environ.* **2015**, *115*, 70–78. [CrossRef]
77. Prasanna, A.; Dorer, V.; Vetterli, N. Optimization of a district energy system with a low temperature network. *Energy* **2017**, *137*, 632–648. [CrossRef]
78. Volkova, A.; Mašatin, V.; Siirde, A. Methodology for evaluating the transition process dynamics towards 4th generation district heating networks. *Energy* **2018**, *150*, 253–261. [CrossRef]

sustainability

MDPI

Review

Methodological Approaches to Optimising Anaerobic Digestion of Water Hyacinth for Energy Efficiency in South Africa

Obianuju Patience Ilo [1,2,*], Mulala Danny Simatele [2], S'phumelele Lucky Nkomo [1], Ntandoyenkosi Malusi Mkhize [3] and Nagendra Gopinath Prabhu [2,4]

1 Discipline of Geography and Environmental Sciences, College of Agriculture, Engineering and Science, University of KwaZulu-Natal, Durban 4041, South Africa; NkomoS1@ukzn.ac.za
2 The Global Change Institute, University of the Witwatersrand, Johannesburg 2050, South Africa; Mulala.Simatele@wits.ac.za (M.D.S.); prabhugn@sdcollege.in (N.G.P.)
3 Discipline of Chemical Engineering, College of Agriculture, Engineering and Science, University of KwaZulu-Natal, Durban 4041, South Africa; MkhizeN7@ukzn.ac.za
4 Centre for Research on Aquatic Resources, S. D. College, University of Kerala, Alleppey 688003, India
* Correspondence: u.jblige@yahoo.com

Citation: Ilo, O.P.; Simatele, M.D.; Nkomo, S.L.; Mkhize, N.M.; Prabhu, N.G. Methodological Approaches to Optimising Anaerobic Digestion of Water Hyacinth for Energy Efficiency in South Africa. *Sustainability* **2021**, *13*, 6746. https://doi.org/10.3390/su13126746

Academic Editors: Luis M. López-Ochoa and José P. Paredes-Sánchez

Received: 5 May 2021
Accepted: 4 June 2021
Published: 15 June 2021

Publisher's Note: MDPI stays neutral with regard to jurisdictional claims in published maps and institutional affiliations.

Abstract: Anaerobic digestion has been identified as a feasible fragment of a bioeconomy, yet numerous factors hinder the adoption of the technology in South Africa. Apart from its energy recovery, other nonmarket advantages support the technology. Though it may be challenging to have a price tag, they provide clear added worth for such investments. With a growing energy demand and global energy transitions, there is a need to sustainably commercialise the biogas industry in South Africa. Most studies are at laboratory scale and under specific conditions, which invariably create gaps in using their data for commercialising the biogas technology. The key to recognising these gaps depends on knowing the crucial technical phases that have the utmost outcome on the economics of biogas production. This study is a meta-analysis of the optimisation of anaerobic digestion through methodological approaches aimed at enhancing the production of biogas. This review, therefore, argues that regulating the fundamental operational parameters, understanding the microbial community's interactions, and modelling the anaerobic processes are vital indicators for improving the process stability and methane yield for the commercialisation of the technology. It further argues that South Africa can exploit water hyacinth as a substrate for a self-sufficient biogas production system in a bid to mitigate the invasive alien plants.

Keywords: biogas; energy transition; water hyacinth; anaerobic digestion; optimisation; sustainable cities

1. Introduction

South Africa's energy is generated predominantly from coal and is rated as the 12th dominant contributors of greenhouse gas (GHG) emissions in the world and the first in Africa [1]. However, attempts at diversifying the energy mix with renewable energy sources such as bioenergy, solar energy, and hydropower are being made, as they are essential contributors to the country's energy supply portfolio and national development. Renewable energy has the potential to contribute to the world energy security and significantly reduce dependency on fossil fuels—oil, coal, and gas [2–4]. They have also provided opportunities for mitigating the devastating effects of greenhouse gases [3,5]. One of the principal renewable energy sources, in the long run, is bioenergy, because it presents a wide range of possibilities such as pyrolysis, anaerobic digestion (AD), and gasification. From a feasibility perspective, each bioenergy technology demonstrates various advantages and disadvantages. For instance, incinerators are frowned upon because of the harmful products such as furans and dioxins released, if not appropriately managed [6,7].

Therefore, consideration for any will be for the availability of feedstock, proximate obtainable infrastructure, and market need. AD has received attention because of its adaptation to a broad spectrum of feedstocks, nutrient recovery, and net energy output [8–10].

AD is described as a biological degradation of organic materials without the presence of oxygen, thereby resulting in methane-rich biogas and an enriched digestate [11]. The methane generated during AD can be converted to different energies depending on indigenous needs; it can be compressed and liquefied as a transportation fuel [12], generation of electricity [13], heat and cooking [2], while the digestate improves the soil structures and reduces the use of chemical fertilisers as shown in Figure 1. AD is considered a flexible technology as its scale of operation can vary from a small size to a much bigger size. It can be integrated with other waste-to-energy technologies such as pyrolysis. AD helps solve societal problems, by creating a sustainable waste management opportunity—mostly by reducing organic wastes on landfills, and economically feasible wastewater treatment.

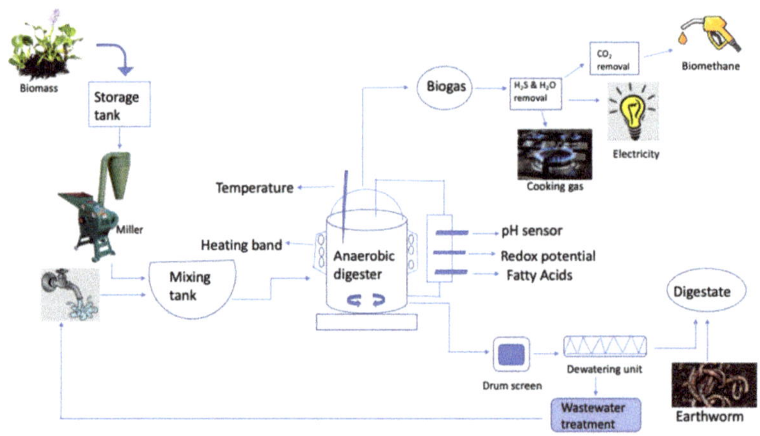

Figure 1. The process flow of anaerobic digestion. Source: Based on personal notes (2021).

In South Africa, AD technology is still considered to be at its inception, even though the technology was first utilised as far back as in the 1950s [14]. The lackadaisical adoption of the technology compared to other countries is ascribed to inadequate feedstocks, lack of institutional support, research output communication, bureaucratic issues such as streamlining the application procedure, and shift of interest to other renewable energy sources [15,16]. According to Tiepelt [17], the number of AD plants in South Africa is not more than 400. Some of the plants are owned by Bio2Watt, WEC/Northern Waste Water Treatment Works, BiogasSA, iBert, and SANEDI. The viability of the biogas industry is largely a function of the cost of feedstock. While most of the deployed biogas plants in the country have concentrated mainly on using animal waste and wastewater as feedstock, there is a broad spectrum of feedstocks and each is dependent on accessibility.

WH (*Eichhornia crassipes*), an invasive alien plant, has gained attention as a promising feedstock for AD because of its high proliferation rate, no threat to food security globally, efficiently hydrolysable sugars, low lignin content, and energy obtained efficiencies. *Eichhornia crassipes* is a native of the Amazon Basin of South America but has spread vastly worldwide. According to Yan and Guo [18], the pervasive occurrence of WH is because of its distinctive biological features, global warming, and intensified eutrophication of surface waters. The prevalence of the aquatic plant in South Africa is attributed to the nutrient enrichment of dams because of the poor treatment of sewage of the heavily populous metropolitans [19]. Its invasion is known to affect sustainability, economic growth, biodiversity, and human health, and the control methods utilised are either not sustainable (chemical), not cost-effective (mechanical), or take a longer time (biological). The expenditure on the control of invasive species in the country is valued to surpass 70 million U.S Dollars annually, which is almost 0.3% of the republic's gross domestic product [20].

WH has been utilised in phytoremediation, compost, animal feed, enzyme production, and bioenergy. Ilo et al. [21] compared the techno-economic feasibility of utilising the aquatic plant to the control methods and reported that utilising *E. crassipes* is economically feasible and sustainable, as most of the cost–benefit analysis for the control methods used models that were based on postulations of market prices. However, the fiscal viability models applied in valuable resource recovery of *E. crassipes* were more genuine and adaptive to likely variations in impending cashflows and discount rates. Therefore, its use in AD is a sustainable method of mitigating its adverse impacts in South Africa. However, the AD process is considered unstable, with a high cost of investment and low return on investment [22]. Optimising the AD process is significant and can considerably add to the decrease of the economic and environmental cost.

This review, therefore, presents WH as a promising and sustainable source of feedstock and aims to provide a comprehensive overview for methodologically improving the efficiency of AD of WH. It further accentuates AD of WH as an economically feasible energy alternative that would sustainably alleviate South Africa's energy crisis. Studies on the effectiveness of a specific methodology recommended by scientists regarding the relationship between the quantity of biogas produced in a laboratory and the possible efficacy when used on an industrial scale are seldom provided [23–25], suggesting a dearth of clarity in methodology. The key to recognising the gaps from laboratory scale to industrial scale of AD depends on knowing the crucial technical phases that have the utmost outcome on the economics [26]. This review consequently describes the strategies for optimising the AD process as (1) regulating the fundamental process parameters, (2) high-throughput molecular tools in understanding the structure of the microbial consortium in AD expose the ecology of unidentified unculturable anaerobic microorganisms, and (3) models and simulations that help comprehend and envisage the AD process for optimisation. Hence, the review envisages that the result of the interpretations would help biogas plant operators improve the operations of AD of WH for optimal energy recovery and help relevant stakeholders understand and facilitate the adoption of the technology in the country.

2. Materials and Methods

This paper is based on a meta-analysis of several empirical works, which aimed at (1) laboratory analysis of AD of WH as both mono- and co-digestion and (2) the process parameters for optimal performance and process stability of AD. An automated search using different databases was executed among which included EBSCOhost, Google Scholar, and Scopus. The following terms were used to search for the relevant studies/and or articles: "biogas", "methane yield", "water hyacinth or *Eichhornia crassipes* "process stability", "co-digestion", "laboratory analysis", "system analysis". The scope of the study was limited to articles published in the past ten years because of the large number of articles retrieved. Furthermore, supplementary investigations were made by studying references of publications and grey literature for more articles that were not retrieved during the search. A total of 1209 records were retrieved and thoroughly examined based on inclusion and exclusion criteria. Excluded articles are (1) articles not available in the English language, (2) unpublished thesis and reports, and (3) articles that are not directly related to the subject areas. The title, abstract, and full-text articles were assessed based on their significance to the objectives stated above; this exercise resulted in the selection of 52 papers. The ROSES flow diagram of the number of studies retrieved is presented in Figure 2. The review made efforts for the search to be all-inclusive.

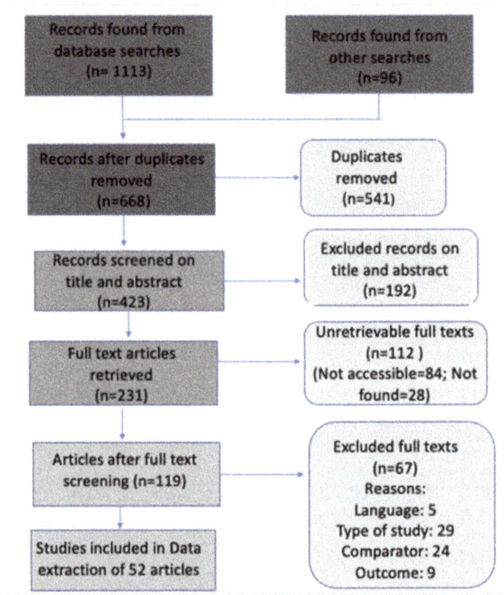

Figure 2. Methodological approach using a rapid appraisal in selected peer-reviewed articles. Source: Adapted from [27].

3. Process Parameters for Optimal AD Performance

Operational and environmental factors are the two sets of parameters that affect AD performance [8]. Attaining an optimum method of AD through regulating the fundamental operational parameters, understanding the microbial community's interactions, and modelling of the anaerobic processes necessitates an assessment of the consequential trade-off between the additional cost of optimisation techniques and improvement in biogas yield. In other words, the basis of optimisation of AD is a favourable relationship between methodological improvements and economic feasibility. While the purpose of improving the technical process comprises optimising biogas production and enhancing process stability, the financial incentive is, therefore, to increase the return on investment. Table 1 represents studies on AD of WH. The studies revealed the effects of pretreatment, mono and co-digestion, temperature, and HRT on biogas and methane yield. This segment presents the methodological review of the AD to highlight the varying biogas yields that are attributed to operational and environmental factors to help scholars greatly comprehend their interactions for optimisation.

Table 1. Studies on AD of WH evaluating numerous process conditions.

S/N	Reference	Digester Type	Substrate	Pretreatment	Temperature (°C)	Inoculum to Substrate Ratio	Methane Yield (%)	Biogas Yield (mL g^{-1} vs)	Hrt (Days)
1	[28]	Batch reactors	WH (Untreated) WH (Pre-treated)	Microbial (*Citrobacter werkmanii* VKVVG4)	-	-	57 ± 0.2 59.99 ± 0.3	156 ± 11	50
2	[29]	Laboratory scale digester	WH + *V. dysplasia* WH+ *P. chrysosporium* WH	Fungal (*Volvariella dysplasia* and *Phanerochaete chrysosporium*)	-	-	64 66 60	99.45 243.66 100	60
3	[30]	ALBR and UASB (20 L)	WH WH + WAS WH + FW	Mechanically crushed	30 ± 3	1:2	63–68	142.8 ± 10 148 ± 5 394.6 ± 12	10
4	[31]	Lab-scale digester	CM + SS + WH	Cut and mashed	37	-	65	81.2	12
5	[32]	Lab-scale digester	WH Salvinia	Blended	37 ± 2	2:1	62 63	552 221	60
6	[33]	Lab-scale anaerobic digester	*Prosopis juliflora* pods + dry leaves + WH	Crushed with blender	35 ± 2	-	47.67 67.66 47.73 42.89	69 96.99 70 62	60
7	[34]	Batch type anaerobic digester (60 L)	FW WH FW + WH (15:2) FW + WH (8:3)	Dried and pulverised	3243	1:2	68.3 65.4 58.2 52.1	370.85 320.54 286.50 298.83	40
8	[35]	Laboratory digester	WH (pretreated)	Cut, blended, and treated with H_2SO_4	28–30	-	64.4	42.40	-
9	[36]	Glass batch reactors	WH + BP (untreated) WH (Pre-treated) + BP	Thermal (Hot air oven)	-	-	57.65 ± 0.2 65.65 ± 0.5	253 ± 3 296 ± 9	50
10	[37]	0.5 L bioreactor vessels	YP + WH, CaP + WH, CoP + WH, PP + WH	Heat dried and milled	37 ± 1	1:2	37.2 23.5 37.8 39.7	419 285.21 387.53 382.46	20

Table 1. *Cont.*

S/N	Reference	Digester Type	Substrate	Pretreatment	Temperature (°C)	Inoculum to Substrate Ratio	Methane Yield (%)	Biogas Yield (mL g^{-1} vs)	Hrt (Days)
11	[38]	Continuous mode two-stage reactor with stage separation (20 L)	WH (untreated) WH (pre-treated) WH + FW	Macerated and preheated WH at 90 °C for 1 hr	35 ± 1	-	57–61 68–71 60–63	-	20
12	[39]	Dry fermentation reactor (5 L)	WH	-	Ambient	1:1	69	41.79	30
13	[40]	2 L Glass batch reactors	WH	-	37 ± 2	0.25:1 0.5:1 1:1 3:1	61 ± 1.5 58 ± 0.33 57 ± 0.67 63 ± 1.4	- 406 - 383	30
14	[41]	500 mL Duran glass bottles	WH WH + Fruit and vegetable waste	-	37	-	57.5 60.5	114 141	15

WH—Water hyacinth, ALBR—Anaerobic Leaching Bed Reactor, UASB—Upflow Anaerobic Sludge Bed Reactor, FW—Food waste, CM—Cow manure, BP—Banana peels, YP—Yam peels, CaP—Cassava peels, CoP—Cocoyam peels, PP—Plantain peels, WAS—Waste Activated Sludge, SS—Sewage sludge.

The biogas yields of the fourteen studies (Table 1) were converted to mL g^{-1} vs for uniformity and ease of comparison. Most of the studies adopted the mesophilic temperature [30–35,37–41] while the rest did not disclose the temperature used [28,29,36]. Furthermore, 35.7% of the studies that used mechanical pretreatment produced a higher biogas yield at the range of 62–552 mL g^{-1} vs [30–34] than those that used other pretreatment processes. Although the chemical process of pretreatment is not popular among the studies, biogas yield was 42.40 mL g^{-1} vs [35]. While 21.4% did not specify the pretreatment used but had biogas yield between 41.79–406 mL g^{-1} vs [39–41], the biological and thermal pretreatment had a biogas yield between 99.45–243.66 mL g^{-1} vs [28,29] and 253 $\pm$ 3–419 mL g^{-1} vs [36–38], respectively. Mathew et al. [32] compared the AD of WH to that of Salvinia and reported a high biogas yield of 552 mL g^{-1} vs to 221 mL g^{-1} vs. The study also revealed a lower volatile fatty acid accumulation (VFA) during the degradation of WH compared to Salvinia.

It can be deduced from Table 1 that co-digestion of WH with other organic materials such as food waste, sewage sludge, and peels produced more methane yield than mono-digestion. In addition, 57.1% of the studies produced more methane yield at an average of 65% than mono digestion at 62%. However, other factors could have either promoted or inhibited the methane and biogas yield in these studies.

3.1. Operational Factors

3.1.1. Effect of Digester Design

The design of a digester is one prominent feature of a cost-effective AD process. The assessment of digester design is dependent on various factors such as cost of installation and maintenance, performance, energy recovery, and discharge of effluents [42]. The different digesters that are commonly constructed are single- or multi-stage, dry or wet, and batch or continuous mode. The single-stage reactor is reported to have fewer technicalities and operational costs; however, the microbial consortiums growth rate is limited as they perform in the same environmental conditions. On the other hand, the multi-stage offers a favourable condition to the microorganisms, but are more complex, necessitate more space, and are not economical [43,44]. Although the multi-stage reactors are considered to improve process stability, it is not factual to state that the single-stage reactors are unreliable. The continuous stirred tank reactor (CSTR) is deemed to be easier to operate than other reactors such as up-flow anaerobic sludge bed reactors (UASB). A digester design should be simple, effective, and economically feasible. Its suitability rating should address parameters such as mixing, temperature, retention time, and the feedstock's quantity and quality, mostly by total solid (TS) basis [45,46].

3.1.2. Mixing

The effects of mixing in AD are to guarantee sufficient access to organic materials for the dynamic microorganisms and to proficiently circulate the heat inside the reactor, thereby inhibiting temperature gradients, dead zones, and hot spots. There are various techniques of mixing, i.e., propellers, recirculation, and each is selected by the type of digester, the TS of the feedstock, and the agitator type. Mechanical mixing has been criticised because of its energy consumption that directly increases operational cost. Mixing is done intermittently so as not to disturb syntrophic activity. Intermittent mixing improves biogas yield because the digestion by-products are degraded better by slower hydrolysis and fermentation [47]. Constant vigorous agitation influences the methane content negatively [48] and the insertion of propellers could trigger an influx of oxygen into the reactor [49]. Of the various mixing techniques, recirculation methods have been proven to be the most economical and efficient in enhancing AD's performance. Ni et al. [50] studied the effect of liquid digestate recirculation on methane yield, and the system balance of AD of pig manure. Under the same operational conditions, the methane yield from the reactors was comparable in phase 1. When the organic loading rate (OLR) was below 5 g vs L^{-1}

in phase 2, there was an increase in Reactor 2 (with recirculation) compared to Reactor 1 (without recirculation), which signified that under comparative OLRs, liquid digestate recirculation stimulated process stability [50].

3.1.3. Hydraulic Retention Time (HRT)

It is an essential factor to consider during the design of a digester as well as for the growth rates of hydrogen- and methane-forming bacteria. It regulates the delivery of substrates to microorganisms. A longer HRT gives the microorganism adequate contact time to degrade the substrates, thereby increasing biogas yield; however, it is considered to increase the operational cost. On the other hand, a shorter HRT is interrelated with volatile fatty acids (VFA) accumulation and washing out of methanogens. The different studies on AD of WH presented in Table 1 were further analysed for the effect of HRT on methane yield and presented in Figure 3. While the analysis illustrates that the highest methane yield of 71–68.3% is at HRT of 20–40 days [34,38,39], nonetheless, at the same HRT of 20 days, the study of Longjan and Dehouche [37] revealed a low methane yield of 23.5–39.7%. A low HRT is considered attractive as it is directly related to a lower investment cost and enhanced process stability [51,52]. The study of Tasnim et al. [31] on co-digestion of cow manure, sewage sludge, and WH revealed a high methane yield of 65% at a 10-day HRT, and Hernández-Shek et al. [41] reported a high methane yield of 60.5% at HRT of 15 days in the co-digestion of WH with fruit and vegetable waste.

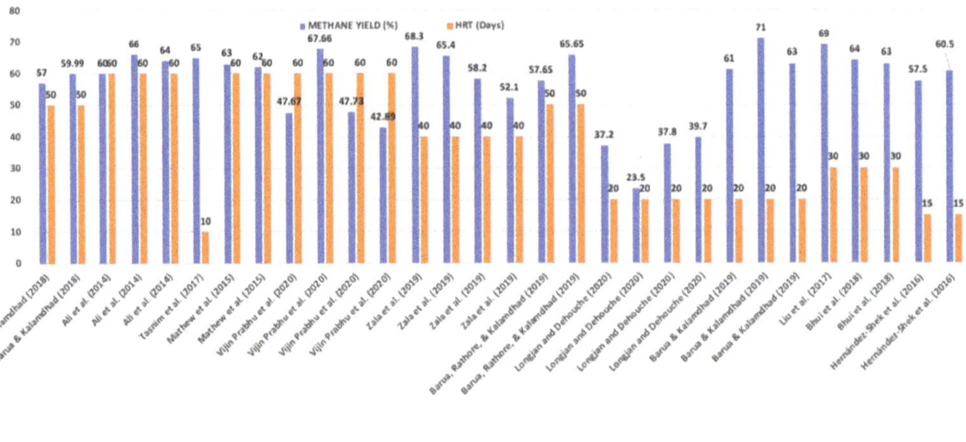

Figure 3. Effect of HRT on methane yield. Source: Based on personal notes (2021).

3.2. *Environmental Factors*

3.2.1. Effect of Inhibition

AD is a delicate process, where the presence of inhibitory compounds such as ammonia and VFA cause system imbalance resulting in low biogas and methane yield [9]. This is because the microbial consortium in each biochemical stage is susceptible to numerous inhibitory matters in the feedstock or produced during the anaerobic process.

Ammonia

Ammonia, which is formed during the biological breakdown of nitrogenous material, is an important function in AD performance and stability. It exists in two basic forms: Ammonium ion (NH_4^+) and free ammonia Nitrogen (NH_3), and a combination of both forms is known as Total Ammonia Nitrogen (TAN). Free ammonia has much higher toxicity and a more significant effect on AD than ammonium ion because it can penetrate cells and disturb microorganisms' metabolism [42,53]. Ideally, ammonia concentration guarantees the methanogenic medium's buffer capacity, but it becomes toxic above a threshold concentration. There are various reports of these threshold levels of toxicity and most do not differentiate between free ammonia and TAN. While Yenigün and Demirel [53] revealed that free ammonia's threshold for toxicity is between 150 to 1200 mg L^{-1}, Rajagopal et al. [54] reported a range of 1500 to 7000 mg L^{-1}.

Studies have shown that methodologies such as pretreatment of feedstocks, co-digestion, air stripping, alteration of pH and temperature, a decrease of OLR, and the addition of support media have been applied in mitigating the inhibitory effect of ammonia [53–55]. Zhang et al. [56], who investigated the effect of air stripping in removing ammonia in AD of piggery wastewater, disclosed that the elimination of ammonia was reliant on pH and aeration rate. Based on their findings, air stripping at alkaline pH is feasible for averting system imbalance in AD of WH.

Volatile Fatty Acids

VFAs are molecular entities formed during the hydrolysis phase, as a result of the degradation of more complex structures. Accumulation of VFA, which causes inhibition, occurs when there is an increase in the OLR. This increase leads to a faster hydrolysis rate that disrupts the acetogen's and methanogens' adaptation, resulting in a drop in pH and a low methane yield [10]. VFA concentrations are used as indicators of process imbalance; however, there are debates on the exact concentration as numerous experiments reveal that process stability occurred at different levels. Mathew et al. [32] reported a total VFA in the AD of WH to be lower than 22 mg L^{-1}. The use of the propionic acid to acetic acid ratio as a sign of process instability is recommended because propionic acids are inhibitory to methanogens [40].

Several approaches can prevent process instabilities as a result of the accumulation of VFA. Rocamora et al. [10] recommended increasing the inoculum: substrate (I:S) ratio and percolate recirculation; however, the study of Bhui et al. [40], which aimed at exploring the effect of VFA in different I:S ratios to biogas production from batch-scale AD of WH, reported maximum VFA accumulations at 1084 mg L^{-1} for WH at 0.25:1 ratio and lowest values at 158 mg L^{-1} for WH at a rate of 3:1. The study revealed that the total VFA drastically affected methane content in 3:1 (WH). Anukam et al. [13] opined for the use of non-biological conductive materials that absorb toxins, which calls for further research.

3.2.2. Organic Loading Rate (OLR)

OLR determines the biogas and methane yield. The kilograms of volatile solids are loaded per volume of the reactor per day [57]. An OLR and the I:S ratio mainly on TS contents simplify the operation process because TS is more realistic than the study of other parameters [47]. Barua and Kalamdhad [38] reported a stable pH when the OLR was increased from 0.625 kg COD m^{3-1} to 1.35 kg COD m^{3-1}; however, the process became unstable when it was further increased to 4.55 kg COD m^{3-1}. The change in pH indicates that the discharge Chemical Oxygen Demand (COD) mainly constitutes the unutilised VFA formed in the reactor at an increased OLR. OLR should be gradual to permit suitable acclimatisation of the microorganism because an abrupt change disturbs them. Nkuna et al. [58] studied the consequences of uneven OLR on microbial communities and AD of WH (mono- and co-digestion) and pointed out that unstable OLR affected the process performance, system balance, and the composition of microorganisms of mono- and co-digestion, but it was dominant in co-digestion.

3.2.3. Temperature

Temperature controls the rates of the enzymatic reaction and substrate diffusion. Most digesters function at either mesophilic (30–40 °C) or thermophilic temperatures (45–60 °C); each temperature has a different active microbial consortium. A slight change in temperature of an anaerobic digester affects the microorganisms' activities, resulting in a low biogas yield. While the thermophilic phase enhances the complex substrate's solubilisation rate, the mesophilic stage provides a stable methane production process [59]. An AD operated in thermophilic temperature has a higher reaction rate, thereby leading to a lesser HRT and digester volumes. It requires high energy for maintaining the reactor at such a high temperature. However, with a slow reaction rate and high HRT, mesophilic digesters are commonly used because of their low energy cost, stable operational process, and less critical ammonium inhibition. Recently, attention has been drawn to the Temperature Phased Anaerobic digesters (TPAD), which involves two-phase systems, both thermophilic and mesophilic phases.

3.2.3.1. pH

pH reveals the approximate condition of a digester, and a drop in pH results in low biogas yield and methanogenesis inhibition. Barua and Kalamdhad [38] noted that a pH of 6.5–7.5 is best for microorganisms to thrive in a digester. However, multi-stage digesters are recommended as the biochemical stages require different optimal pH values. At the same time, Rocamora et al. [10] reported an optimal pH of hydrolysis and acidogenesis to be within the range of 5.5 and 6.5, and Mao et al. [60] opined that methanogenesis occurs at a higher pH of 6.5 and 8.2, with an optimum at 7.0.

Studies have shown that it is not ideal to use pH as a first pointer for process stability because it relies on buffering capacity. For instance, Yi et al. [61] analysed the role of increasing total solids on the performance of AD of food waste at mesophilic temperature. Digesters with higher TS had higher VFAs accumulation, but digester R3, which had the highest VFAs accumulation, did not indicate a low pH. The study of Widyarani et al. [62] on the effect of pH on biogas generation of tofu wastewater revealed that low pH did not negatively affect the batch AD of the Tofu wastewater system. The outcome of their study implies that ensuring there is buffer capacity is imperative in comparison to adjusting pH.

3.2.4. Co-Digestion

The anaerobic mono-digestion of *E. crassipes* is rate-limiting because the hydrolysis of the lignocellulosic structure takes a long time and reduces biogas yield [36]. Co-digestion is an efficient and commercially feasible method to improve methane production and system stability [63]. The essence of co-digestion is to adjust the carbon/nitrogen (C/N) ratio of the feedstocks for efficient microbial growth. C/N ratio shows the nutrient levels of feedstocks. A high C/N ratio results in a shortage of nitrogen and dormancy of methanogens, reducing methane yield. In contrast, a low C/N ratio leads to carbon shortage, thereby causing the accumulation of VFA, which has a negative effect during methanogenesis.

Priya et al. [30] tested practical solutions to enhance biogas production from the AD of WH. They reported a biogas yield from the co-digestion of WH, with activated sludge and food waste as ~150 mL g^{-1} vs and ~400 mL g^{-1} vs, respectively. In contrast, mono-digestion of WH yielded ~140 mL g^{-1} vs of biogas. However, Zala et al. [34] reported a higher biogas yield of 320.5 mL g^{-1} vs in mono-digestion of WH compared to the biogas yield of 286.5 mL g^{-1} vs and 298.8 mL g^{-1} vs in co-digestion of WH and food waste, but the difference in pH rate displayed process stability in co-digestion than in mono-digestion.

3.2.5. FOS/TAC

Several parameters such as methane yield, pH, or VFA are used as indicators for process stability in AD; however, the FOS/TAC is extensively reflected as the most imperative and express marker [64]. The titration method, which represents the ratio between volatile organic acids (FOS) and total inorganic carbonate (buffer capacity) (TAC), is an easy and continuous method for determining AD process stability [58,65]. The composition of feedstock for AD affects the FOS/TAC value. Nkuna and Roopnarain [58] reported process stability and high biogas yield from the AD of WH at FOS/TAC ratio of 0.4–0.6. A high FOS/TAC value (>0.6) implies system overload, and this results in a low yield of methane, whereas a low FOS/TAC value indicates a low OLR, which causes process imbalance.

3.3. *Other Factors*

3.3.1. Pretreatment

Production of biogas from lignocellulosic biomass is demanding because its structure and composition makes it recalcitrant to microbial or enzymatic degradation. However, pretreatment is an efficient approach to breaking down the organic molecules' covalent bonds, decreasing the recalcitrance, and increasing biodegradability. It exposes the lignocellulosic polymers into hexoses and pentoses, helps microorganisms access the cellulose, and hastens the hydrolysis stage. There are different pretreatment methods; however, the choice for pre-treatment should be sustainable, cost-effective, and not yield inhibitory compounds.

A lack of empirical consensus still exists in the literature over an established single pretreatment method that is the most effective for high methane yield. Sarto et al. [35] examined the effect of chemical pretreatment (H_2SO_4) in facilitating the production of biogas from WH, and although the cellulose content was broken down significantly to glucose, the lignin composition slightly decreased. Barua et al. [28] investigated the effect of microbial (*Citrobacter werkmanii* VKVVG4) pretreatment on WH. The study showed that, although microbial pretreatment consumed time to improve the solubility and breakdown of WH's lignocellulosic cell wall, it enhanced the biogas yield. In recent time, the use of integrated pretreatment methods has been utilised for optimal methane production.

3.3.2. Inoculum

Inoculum with the balanced microbial consortium is an important factor that reduces the acclimatisation period for process stability and efficiency of AD performance [66]. It is a proficient method of delivering the essential microorganisms to the AD process. In recent times, specific microorganisms are used as inoculants to increase degradation rate, unlike previously, where indigenous microorganisms conducted the degradation. The presence of key members of the anaerobic microbial consortium in an inoculum strongly influences the AD process's performance [38]. A suitable I/S ratio circumvents process instability in a digester by creating a conducive atmosphere for microbial activities. A high I/S ratio enhances the efficiency of removing COD as it hastens COD's breakdown to biogas [40]. COD is a suitable indicator that reveals the extent the degradation process has taken to be completed; the higher the COD removal, the more stable the process is.

Examining how microbial consortium change at the start-up phase increases the microbial community's relationship to AD performance, thereby improving the process economics. Studies have engaged in using high-throughput methodologies such as 454 pyrosequencing, quantitative Polymerase Chain Reaction (qPCR), and fluorescent in situ hybridisation (FISH) for such investigations [67,68]. The use of support media (immobilisers) stimulates methanogenic reactions by creating opportunities on feedstock's surface area for microbial attachment. Although there is limited literature on the consequences of these support media to the microorganisms, the study of Poirier et al. [69] emphasised that immobilisers such as zeolites and activated carbon aided the AD process under high ammonium stress and also improved the growth of the microorganisms.

4. Modelling of AD Process

Although it is imperative to acquire knowledge from conducting tests, a theoretical evaluation must propose a hypothesis and establish a linkage for knowing and optimising the AD process. According to Kucharska et al. [70], modelling is used in optimising the process parameters, thereby saving time and increasing the efficacy of utilising resources. Mathematical models aid in investigating phenomena during the AD process; they are also used to transfer experiments from pilot-scale to industrial scale. International Water Associations' Anaerobic Digestion Model No 1 (ADMI) is the most generally known model in AD studies; however, its intricacy had steered the inclusion of the latest empirical understanding or simpler algorithms [71]. Other modeling approaches utilised on the AD optimisation process are classified by Kucharska et al. [70] as Kinetic models such as

Gompertz and Monod models; black-box models such as Response Surface Methodology (RSM) and Artificial Neural Networks (ANN) and Substrate conversion-based models such as First-order. Logistic and Boltzmann [72] and AD model No. 2 (AM2) [71] have also been applied for fitness.

The black-box models predict complicated processes with unspecified input and output data correlation. It does not require an antecedent understanding of the mechanism. The model is applied when a reaction is influenced by numerous variables; it assesses the relationship between biogas yield and independent parameters. ANN is considered to have a higher prediction capability that is close to measured gases compared to RSM [70]. Chanathaworn [73] improved the biogas production from the AD of WH and earthworm bedding wastewater using the RSM model approach, and TS, pH, and particle size as the model variables. The study confirmed the fitness of the model at a coefficient of determination (R^2) of 96.1%

The kinetic models consider microbial consortium as an essential component of the digester. For instance, the modified Gompertz equation reveals the correlation between methane yield and microbial growth pattern [28]. On the other hand, the substrate conversion-based models aim at the substate decomposition or biogas yield and not on microbial growth rate. For instance, the First-order Kinetics provides changes in volatile solids during biodegradation, although it is considered to have an extensive latency period [74] and not predict process failure [75].

The R^2 and root of the mean of the squares (RMSE) are usually used to compare fitting errors of models. Sarto et al. [35] compared the simulation of biogas production from *E. crassipes* with a modified Gompertz, First-order kinetics, and Cone model, and reported that all the proposed models fitted the determined biogas yield with fitting inaccuracy that is below 10%. The variation between determined and theorised biogas yield for modified Gompertz was 0.271–9.78%, 3.491–5.424% for First-order Kinetic, and 0.032–8.743% for the Cone model, while the R^2 was 0.964–0.995 (modified Gompertz), 0.977–0.985 (First-order Kinetic), and 0.980–0.994 (Cone), respectively.

5. Outlook of Biogas Technology in South Africa's Energy Transition

The Energy sector of South Africa contributed to nearly 80% of the nation's Greenhouse gas (GHG) emissions in 2019, of which half originated from petrol production and power generation [76]. Consequently, South Africa has pledged in the Integrated Resource Plan 2019 to a low-carbon economy to address climate change. However, there is a need for this transition to be timely as the country is facing the menace of climate change. The present COVID-19 pandemic is an avenue for the energy transition to achieve the climate targets and promote socio-economic development [77–79]. The three priority areas for energy transition for sustainable cities are buildings, transport, and integrated energy systems; this makes AD an integral technology for South Africa's bioeconomy because it is multifaceted.

The use of biomethane in the transport sector is promising, though it is still at the infant level in the country. Biomethane can be compressed and utilised in passenger vehicles or liquefied and used in heavy-haul vehicles and ships. Its utilisation as a transport fuel is considered to have limited environmental concerns, such as GHG emissions, compared to fossil-derived fuels [12,80]. In addition, compared to other biofuels, biomethane is unique in blending wholly with natural gas without requiring the engine to be changed. For example, the blending of bioethanol with petrol is usually at portions of 5, 10, and 85% quantities, and while the portions of 5 and 10 do not require amendment of vehicle, the use of an 85% ratio does [81]. This amendment of the vehicle (Flexi-Fuel vehicles) is reported to possibly release unburned ethanol, acetaldehyde, and acetic acid in the environment [82] and also gives rise to disastrous engine failure because the bioethanol is corrosive and it forms films as it is mixable in water and non-mixable in oil [79,83].

A fundamental question is whether the numerous valuable functions linked to the AD technology can be significantly realistic and sustainable. One of the constraining facets of

AD is the availability of feedstocks. While the energy crops are frowned at as unsustainable, agricultural residues and organic fraction of municipal solid waste are gaining attention. The lignocellulosic feedstock trade is liable to develop quickly in the long run when compared to conventional feedstocks. Other limitations of the biogas industry's growth ultimately rely on sound policy and institutional framework, market players, and technical expertise. In the global energy market, the biofuel markets are in poor trade patterns against fossil fuel markets notwithstanding the tariffs and no-tariff trade difficulties. The South African government had introduced a renewable energy feed-in-tariffs (REFiTs) program, which had poor implementation and was substituted by an auction-based tariff under the Renewable Energy Independent Power Producer Procurement (REIPP) program [84]. Its sustainability is questionable as it is considered to have an expensive transaction fee, and the monitoring and assessment procedure are not accessible to the public. With the growing energy demand in the country, there is a need to scale up the biogas industry, which, even now, has not been adopted fully; however, one needs to certify the production approach and the kind of feedstocks used.

6. Conclusions

High energy demand and accessibility of ample feedstocks have made AD a promising business opportunity. The AD technology decreases carbon emissions, offers energy security, and creates green jobs. Commercialisation of biogas production has not been adopted to its full potential in South Africa due to sound policy and institutional framework, a market economy, technical expertise, and sustainable feedstocks. While research and investments in biogas production are strongly recommended for upscaling biogas production to a commercial level, there are, however, substantial gaps in the literature due to disagreements over the most economical methods or combination of methods that generates optimal biogas yield. Attaining an optimum method of AD of WH through regulating the fundamental operational parameters, understanding the microbial community's interactions, and modelling of the anaerobic processes necessitates an assessment of the consequential trade-off between the additional cost of optimisation techniques and improvement in biogas yield. It was deduced from the studies that co-digestion of WH and other organic materials such as food waste and sewage sludge produced more methane yield of 65% than mono digestion of 62%. Furthermore, 35.7% of the studies that used mechanical pretreatment produced a higher biogas yield at the range of 62–552 mL g^{-1} vs than those that used other pretreatment processes. These methods could be extended to the utilisation of other feedstocks notwithstanding that the present review focused on WH as they share operational similarities. It is envisaged that the thematic issues addressed in this paper will contribute to policy discourse and scholarly deliberations that would engender more research and investments towards upscaling biogas production to the commercial phase.

Author Contributions: Writing—original draft preparation, O.P.I.; methodology, O.P.I.; writing—review and editing, S.L.N., N.M.M., N.G.P. and M.D.S.; supervision, S.L.N. and N.M.M. All authors have read and agreed to the published version of the manuscript.

Funding: This research received no external funding.

Institutional Review Board Statement: Not applicable.

Informed Consent Statement: Not applicable.

Data Availability Statement: No new data were created in this study.

Conflicts of Interest: The authors declare no conflict of interest.

Abbreviations

AD—Anaerobic Digestion, WH—Water hyacinth, LCFA—Long-chain fatty acids, CSTR—Continuous stirred tank reactor, ALBR—Anaerobic leaching bed reactor, UASB—Upflow anaerobic sludge bed reactor, TPAD—Temperature phased anaerobic digester, VS—Volatile Solids, FW—Food waste, HRT—Hydraulic retention time, C/N—Carbon-Nitrogen ratio, VFA—Volatile fatty acids, I/S—Inoculum-Substrate ratio, TS—Total solids, CH_4—Methane, OLR—Organic loading rate, TAN—Total Ammonia Nitrogen, COD—Chemical Oxygen Demand, GHG—Greenhouse gas.

References

1. Bloomberg Green. South Africa Plans 28% Lower 2030 Carbon Emissions Target. Available online: https://www.bloomberg.com/news/articles/2021-03-30/south-africa-speeds-up-plans-to-reduce-carbon-emissions (accessed on 19 May 2021).
2. Owusu, P.A.; Asumadu-Sarkodie, S.A. Review of renewable energy sources, sustainability issues and climate change mitigation. *Cogent Eng.* **2016**, *3*, 1–14. [CrossRef]
3. Gielen, D.; Boshell, F.; Saygin, D.; Bazilian, M.D.; Wagner, N.; Gorini, R. The role of renewable energy in the global energy transformation. *Energy Strategy Rev.* **2019**, *24*, 38–50. [CrossRef]
4. Kåberger, T. Progress of renewable electricity replacing fossil fuels. *Global Energy Interconnect.* **2018**, *1*, 48–52. [CrossRef]
5. IPCC. Summary for policymakers. In *IPCC Special Report on Renewable Energy Sources and Climate Change Mitigation*; Edenhofer, O., Pichs-Madruga, R., Sokona, Y., Seyboth, K., Matschoss, P., Kadner, S., Zwickel, T., Eickemeier, P., Hensen, G., Schlömer, S., et al., Eds.; Cambridge University Press: Cambridge, UK, 2011; pp. 1–230.
6. Mukherjee, A.; Debnath, B.; Ghosh, S.K. A Review on technologies of removal of dioxins and furans from incinerator flue gas. *Procedia Environ. Sci.* **2016**, *35*, 528–540. [CrossRef]
7. Nzihou, A.; Themelis, N.J.; Kemiha, M.; Benhamou, Y. Dioxin emissions from municipal solid waste incinerators (MSWIs) in France. *Waste Manag.* **2012**, *32*, 2273–2277. [CrossRef] [PubMed]
8. Rabii, A.; Aldin, S.; Dahman, Y.; Elbeshbishy, E. A Review on anaerobic co-digestion with a focus on the microbial populations and the effect of multi-stage digester configuration. *Energies* **2019**, *12*, 1106. [CrossRef]
9. Kumar, A.; Samadder, S.R. Performance evaluation of anaerobic digestion technology for energy recovery from organic fraction of municipal solid waste: A review. *Energy* **2020**, *197*. [CrossRef]
10. Rocamora, I.; Wagland, S.T.; Villa, R.; Simpson, E.W.; Fernández, O.; Bajón-Fernández, Y. Dry anaerobic digestion of organic waste: A review of operational parameters and their impact on process performance. *Bioresour. Technol.* **2020**, *299*, 1–11. [CrossRef]
11. Naik, L.; Gebreegziabher, Z.; Tumwesige, V.; Balana, B.B.; Mwirigi, J.; Austin, G. Factors determining the stability and productivity of small scale anaerobic digesters. *Biomass Bioenergy* **2014**, *70*, 51–57. [CrossRef]
12. Dahlgren, S. Biogas-based fuels as renewable energy in the transport sector: An overview of the potential of using CBG, LBG and other vehicle fuels produced from biogas. *Biofuels* **2020**, 1–13. [CrossRef]
13. Anukam, A.; Mohammadi, A.; Naqvi, M.; Granström, K. A review of the chemistry of anaerobic digestion: Methods of accelerating and optimizing process efficiency. *Processes* **2019**, *7*, 504. [CrossRef]
14. GIZ.; SALGA. Biogas to Energy at Municipal Wastewater Treatment Works: A Toolkit for Municipalities to Assess the Potential at Individual Plants. Deutsche Gesellschaft fur International Zusammenarbeit (GIZ) GmbH, South Africa, 2015. Available online: https://cityenergy.org.za/uploads/resource_327.pdf (accessed on 19 May 2021).
15. Roopnarain, A.; Adeleke, R. Current status, hurdles and future prospects of biogas digestion technology in Africa. *Renew. Sustain. Energy Rev.* **2017**, *67*, 1162–1179. [CrossRef]
16. Mukumba, P.; Makaka, G.; Mamphweli, S. Biogas technology in South Africa, problems, challenges and solutions. *Int. Journal Sustain Energy Environ. Res.* **2016**, *5*, 58–69. [CrossRef]
17. Tiepelt, M. Status quo of the biogas sector development in South Africa as well as the way forward. In Proceedings of the GIZ SAGEN Short-Term Biogas Training Seminar, Pretoria, South Africa, 10 June 2015.
18. Yan, S.H.; Guo, J.Y. *Water hyacinth: Environmental Challenges, Management and Utilization*; CRC Press, Taylor and Francis Group, LLC: Boca Raton, FL, USA, 2017; pp. 1–345. ISBN 13:978-1-4987-4389-1.
19. Auchterlonie, J.; Eden, C.L.; Sheridan, C. The phytoremediation potential of water hyacinth: A case study from Hartbeespoort Dam, South Africa. *S. Afr. J. Chem. Eng.* **2021**, *37*, 31–36. [CrossRef]
20. Van Wilgen, B.W.; de Lange, W.J. The Costs and Benefits of Biological Control of Invasive Alien Plants in South Africa. *Afr. Entomol.* **2011**, *19*, 504–514. [CrossRef]
21. Ilo, O.P.; Simatele, M.D.; Nkomo, S.L.; Mkhize, N.M.; Prabhu, N.G. The benefits of water hyacinth (*Eichhornia crassipes*) for Southern Africa: A review. *Sustainability* **2020**, *12*, 9222. [CrossRef]
22. Winquist, E.; Rikkonen, P.; Pyysiäinen, J.; Varho, V. Is biogas an energy or a sustainability product? Business Opportunities in the Finnish biogas branch. *J. Clean. Prod.* **2019**, *233*, 1344–1354. [CrossRef]
23. Bouallagui, H.; Marouani, L.; Hamdi, M. Performances comparison between laboratory and full-scale anaerobic digesters treating a mixture of primary and waste activated sludge. *Resour. Conserv. Recycl.* **2010**, *55*, 29–33. [CrossRef]
24. Hamawand, I.; Baillie, C. Anaerobic digestion and biogas potential: Simulation of lab and industrial-scale processes. *Energies* **2015**, *8*, 454–474. [CrossRef]
25. Pilarski, K.; Pilarska, A.A.; Boniecki, P.; Niedbała, G.; Durczak, K.; Witaszek, K.; Mioduszewska, N.; Kowalik, I. The efficiency of industrial and laboratory anaerobic digesters of organic substrates: The use of the biochemical methane potential correction coefficient. *Energies* **2020**, *13*, 1280. [CrossRef]

26. Achinas, S.; Achinas, V.; Euverink, G.J.W. A technological overview of biogas production from biowaste. *Engineering* **2017**, *3*, 299–307. [CrossRef]
27. Haddaway, N.R.; Macura, B.; Whaley, P.S.B.; Pullin, A.S. ROSES Flow Diagram for Systematic Reviews 2017. Available online: https://www.roses-reporting.com/flow-diagram (accessed on 3 March 2021).
28. Barua, V.B.; Goud, V.V.; Kalamdhad, A.S. Microbial pretreatment of water hyacinth for enhanced hydrolysis followed by biogas production. *Renew. Energy* **2018**, *126*, 21–29. [CrossRef]
29. Ali, N.; Chaudhary, B.L.; Panwar, N.L. The fungal pre-treatment of maize cob heart and water hyacinth for enhanced biomethanation. *Int. J. Green Energy* **2014**, *11*, 40–49. [CrossRef]
30. Priya, P.; Nikhitha, S.O.; Anand, C.; Dipin Nath, R.S.; Krishnakumar, B. Biomethanation of water hyacinth biomass. *Bioresour. Technol.* **2018**, *255*, 288–292. [CrossRef]
31. Tasnim, F.; Iqbal, S.A.; Chowdhury, A.R. Biogas production from anaerobic co-digestion of cow manure with kitchen waste and Water Hyacinth. *Renew. Energy* **2017**, *109*, 434–439. [CrossRef]
32. Mathew, A.; Bhui, I.; Banerjee, S.N.; Goswami, R.; Chakraborty, A.K.; Shome, A.; Balachandran, S.; Chaudhury, S. Biogas production from locally available aquatic weeds of Santiniketan through anaerobic digestion. *Clean Technol. Environ. Policy* **2015**, *17*, 1681–1688. [CrossRef]
33. Vijin Prabhu, A.; Manimaran, R.; Antony Raja, S.; Jeba, P. Biogas production from anaerobic co-digestion of Prosopis juliflora pods with water hyacinth, dry leaves, and cow manure. *Energy Sources Part A Recovery Util. Environ. Eff.* **2020**, *42*, 375–386. [CrossRef]
34. Zala, M.; Ravi, S.; Bhale, P.V.; Vaishak, S. Experimental investigation on anaerobic co-digestion of food waste and water hyacinth in batch type reactor under mesophilic condition. *Biomass Convers. Biorefinery* **2019**, *10*, 707–714. [CrossRef]
35. Sarto, S.; Hildayati, R.; Syaichurrozi, I. Effect of chemical pretreatment using sulfuric acid on biogas production from water hyacinth and kinetics. *Renew. Energy* **2019**, *132*, 335–350. [CrossRef]
36. Barua, V.B.; Rathore, V.; Kalamdhad, A.S. Anaerobic co-digestion of water hyacinth and banana peels with and without thermal pretreatment. *Renew. Energy* **2019**, *134*, 103–112. [CrossRef]
37. Longjan, G.G.; Dehouche, Z. Biogas production potential of co-digested food waste and water hyacinth common to the Niger Delta. *Biofuels* **2017**, *11*, 277–287. [CrossRef]
38. Barua, V.B.; Kalamdhad, A.S. Biogas production from water hyacinth in a novel anaerobic digester: A continuous study. *Process Saf. Environ. Prot.* **2019**, *127*, 82–89. [CrossRef]
39. Liu, J.; Yan, L.; Wang, Q.; Zhao, X.; Wang, C.; Liu, J.; Yang, H.; Yin, F.; Zhange, W.-D. Experimental study on dry fermentation of water hyacinth residue by anaerobic digestion. In Proceedings of the 2016 International Conference on Energy Development and Environmental Protection, Beijing, China, 11–12 June 2016.
40. Bhui, I.; Mathew, A.K.; Chaudhury, S.; Balachandran, S. Influence of volatile fatty acids in different inoculum to substrate ratio and enhancement of biogas production using water hyacinth and Salvinia. *Bioresour. Technol. Rep.* **2018**, *270*, 409–415. [CrossRef] [PubMed]
41. Hernández-Shek, M.A.; Cadavid-Rodríguez, L.S.; Bolaños, I.V.; Agudelo-Henao, A.C. Recovering biomethane and nutrients from anaerobic digestion of water hyacinth (*Eichhornia crassipes*) and its co-digestion with fruit and vegetable waste. *Water Sci. Technol.* **2016**, *73*, 355–361. [CrossRef] [PubMed]
42. Mir, M.A.; Hussain, A.; Verma, C.; Dubey, S. Design considerations and operational performance of anaerobic digester: A review. *Cogent Engineering* **2016**, *3*. [CrossRef]
43. Pham Van, D.; Takeshi, F.; Hoang Minh, G.; Pham Phu, S.T. Comparison between single and two-stage anaerobic digestion of vegetable waste: Kinetics of methanogenesis and carbon flow. *Waste Biomass Valorization* **2019**, *11*, 6095–6103. [CrossRef]
44. Hernandez, J.E.; Edyvean, R.G.J. Comparison between a Two-Stage and Single-Stage Digesters When Treating a Synthetic Wastewater Contaminated with Phenol. *Water SA* **2011**, *37*, 27–32. Available online: http://www.scielo.org.za/pdf/wsa/v37n1/v37n1a05.pdf (accessed on 13 May 2021). [CrossRef]
45. Krishania, M.; Kumar, V.; Vijay, V.K.; Malik, A. Analysis of different techniques used for improvement of biomethanation process: A review. *Fuel* **2013**, *106*, 1–9. [CrossRef]
46. Baltrėnas, P.; Baltrėnaitė, E. The types, structure and design of small-scale bioreactors producing biogas. In *Small Bioreactors for Management of Biodegradable Waste*; Springer International Publishing: Cham, Switzerland, 2018; pp. 91–149. [CrossRef]
47. Li, C.; Champagne, P.; Anderson, B.C. Biogas production performance of mesophilic and thermophilic anaerobic co-digestion with fat, oil, and grease in semi-continuous flow digesters: Effects of temperature, hydraulic retention time, and organic loading rate. *Environ Technol.* **2013**, *34*, 2125–2133. [CrossRef] [PubMed]
48. Singh, B.; Szamosi, Z.; Simenfalvi, Z. Impact of mixing intensity and duration on biogas production in an anaerobic digester: A review. *Crit Rev Biotechnol.* **2020**, *40*, 508–521. [CrossRef]
49. Narayanan, C.M.; Narayan, V. Biological wastewater treatment and bioreactor design: A review. *Sustain. Environ. Res.* **2019**, *29*. [CrossRef]
50. Ni, P.; Lyu, T.; Sun, H.; Dong, R.; Wu, S. Liquid digestate recycled utilization in anaerobic digestion of pig manure: Effect on methane production, system stability and heavy metal mobilization. *Energy* **2017**, *141*, 1695–1704. [CrossRef]
51. Ezekoye, V.A.; Ezekoye, B.A.; Offor, P.O. Effect of retention time on biogas production from poultry droppings and cassava peels. *Niger. J. Biotechnol.* **2011**, *22*, 53–59.

52. Shi, X.S.; Dong, J.J.; Yu, J.H.; Yin, H.; Hu, S.M.; Huang, S.X.; Yuan, X.Z. Effect of hydraulic retention time on anaerobic digestion of wheat straw in the semicontinous continuous stirred-tank reactors. *BioMed Res. Int.* **2017**, 1–7. [CrossRef]
53. Yenigün, O.; Demirel, B. Ammonia inhibition in anaerobic digestion: A review. *Process Biochem.* **2013**, *48*, 901–911. [CrossRef]
54. Rajagopal, R.; Massé, D.I.; Singh, G. A critical review on inhibition of anaerobic digestion process by excess ammonia. *Bioresour. Technol.* **2013**, *143*, 632–641. [CrossRef] [PubMed]
55. Jiang, Y.; McAdam, E.; Zhang, Y.; Heaven, S.; Bank, C.; Longhurst, P. Ammonia inhibition and toxicity in anaerobic digestion: A critical review. *J. Water Process Eng.* **2019**, *32*. [CrossRef]
56. Zhang, L.; Lee, Y.W.; Jahng, D. Ammonia stripping for enhanced biomethanization of piggery wastewater. *J. Hazard. Mater.* **2012**, *199–200*, 36–42. [CrossRef]
57. Fagbohungbe, M.O.; Dodd, I.C.; Herbert, B.M.J.; Li, H.; Ricketts, L.; Semple, K.T. High solid anaerobic digestion: Operational challenges and possibilities. *Environ. Technol. Innov.* **2015**, *4*, 268–284. [CrossRef]
58. Nkuna, R.; Roopnarain, A.; Adeleke, R. Effects of organic loading rates on microbial communities and biogas production from water hyacinth: A case of mono- and co-digestion. *J. Chem. Technol. Biotechnol.* **2019**, *94*, 1294–1304. [CrossRef]
59. Jideofor, O.P.; Akunna, J.; Onyia, C.O.; Solomon, B.O. A Comparative Study of the Performance of Anaerobic Digestion of Seaweed (*Ascophyllum nodosum*) At Mesophilic and Thermophilic Temperatures. *Global J. Biol. Agric. Health Sci.* **2013**, *2*, 55–58.
60. Mao, C.; Feng, Y.; Wang, X.; Ren, G. Review on research achievements of biogas from anaerobic digestion. *Renew. Sustain. Energy Rev.* **2015**, *45*, 540–555. [CrossRef]
61. Yi, J.; Dong, B.; Jin, J.; Dai, X. Effect of increasing total solids contents on anaerobic digestion of food waste under mesophilic conditions: Performance and microbial characteristics analysis. *PLoS ONE* **2014**, *9*, e102548. [CrossRef]
62. Widyarani, W.; Victor, Y.; Sriwuryandari, L.; Priantoro, E.A.; Sembiring, T.; Sintawardani, N. Influence of pH on biogas production in a batch anaerobic process of tofu wastewater. *IOP Conf. Ser. Earth Environ. Sci.* **2018**, *160*. [CrossRef]
63. Jegede, A.O.; Zeeman, G.; Bruning, H. A review of mixing, design and loading conditions in household anaerobic digesters. *Crit. Rev. Environ. Sci. Technol.* **2019**, *49*, 2117–2153. [CrossRef]
64. Costa, A.; Maria Tangorra, F.; Zaninelli, M.; Oberti, R.; Guidobono Cavalchini, A.; Savoini, G.; Lazzari, M. Evaluating an e-nose ability to detect biogas plant efficiency: A case study. *Ital. J. Anim. Sci.* **2016**, *15*, 116–123. [CrossRef]
65. Lili, M.; Biró, G.; Sulyok, E.; Patis, M.; Janos, B.; Janos, T. Novel approach of the basis of FOS / TAC method. In Proceedings of the International Symposia "Risk Factors for Environment and Food Safety" & "Natural Resources and Sustainable Development" & "50 Years of Agriculture Researche in Oradea", University of Oradea, Faculty of Environmental Protection, Oradea, Romania, 4–5 November 2011; pp. 802–807.
66. Kainthola, J.; Kalamdhad, A.S.; Goud, V.V. Optimization of process parameters for accelerated methane yield from anaerobic co-digestion of rice straw and food waste. *Renew. Energy* **2020**, *149*, 1352–1359. [CrossRef]
67. Wu, B.; Wang, X.; Deng, Y.-Y.; He, X.-L.; Li, Z.-W.; Li, Q.; Qin, H.; Chen, J.-T.; He, M.-X.; Zhang, M.; et al. Adaption of microbial community during the start-up stage of a thermophilic anaerobic digester treating food waste. *Biosci. Biotechnol. Biochem.* **2016**, *80*, 2025–2032. [CrossRef] [PubMed]
68. Liu, T.; Sun, L.; Muller, B.; Schnurer, A. Importance of inoculum source and initial community structure for biogas production from agricultural substrates. *Bioresour. Technol.* **2017**, *245 Pt A*, 768–777. [CrossRef]
69. Poirier, S.; Madigou, C.; Bouchez, T.; Chapleur, O. Improving anaerobic digestion with support media: Mitigation of ammonia inhibition and effect on microbial communities. *Bioresour. Technol.* **2017**, *235*, 229–239. [CrossRef]
70. Kucharska, K.; Hołowacz, I.; Konopacka-Łyskawa, D.; Rybarczyk, P.; Kamiński, M. Key issues in modeling and optimization of lignocellulosic biomass fermentative conversion to gaseous biofuels. *Renew. Energy* **2018**, *129*, 384–408. [CrossRef]
71. Arzate, J.A.; Kirstein, M.; Ertem, F.C.; Kielhorn, E.; Malule, H.R.; Neubauer, P.; Cruz-Bournazou, M.N.; Junne, S. Anaerobic digestion model (AM2) for the description of biogas processes at dynamic feedstock loading rates. *Chem. Ing. Tech.* **2017**, *89*, 686–695. [CrossRef]
72. Teleky, B.E.; Bălan, M.C.; Nikolausz, M. Timeline of Bio-Hydrogen Production by Anaerobic Digestion of Biomass. *Leonardo Electron. J. Pract. Technol.* **2015**, *27*, 67–80. Available online: http://lejpt.academicdirect.ro/A27/067_080.pdf (accessed on 18 May 2021).
73. Chanathaworn, J. Operating condition optimization of water hyacinth and earthworm bedding wastewater for biogas production. *Energy Procedia* **2017**, *138*, 253–259. [CrossRef]
74. Zhang, H.; Luo, L.; Li, W.; Wang, X.; Sun, Y.; Sun, Y.; Gong, W. Optimization of mixing ratio of ammoniated rice straw and food waste co-digestion and impact of trace element supplementation on biogas production. *J. Mater. Cycles Waste Manag.* **2018**, *20*, 745–753. [CrossRef]
75. Kafle, G.K.; Chen, L. Comparison on batch anaerobic digestion of five different livestock manures and prediction of biochemical methane potential (BMP) using different statistical models. *Waste Manag.* **2016**, *48*, 492–502. [CrossRef]
76. Vuppuluri, P. South Africa's Energy Crisis: Three Things to Consider in the Transition toward Renewable Energy. 2020. Available online: https://www.forbes.com/sites/forbesfinancecouncil/2020/10/05/south-africas-energy-crisis-three-things-to-consider-in-the-transition-toward-renewable-energy/?sh=743859e03f89 (accessed on 28 April 2021).
77. UNEP. Transforming the Energy System—A Post-COVID-19 win-Win for People and Planet. 2020. Available online: https://www.unep.org/news-and-stories/story/transforming-energy-system-post-covid-19-win-win-people-and-planet (accessed on 4 March 2021).

78. Kuzemko, C.; Bradshaw, M.; Bridge, G.; Goldthau, A.; Jewell, J.; Overland, I.; Scholten, D.; van de Graaf, T.; Westphal, K. Covid-19 and the politics of sustainable energy transitions. *Energy Res. Social Sci.* **2020**, *68*, 1–7. [CrossRef] [PubMed]
79. Gebreslassie, M.G. Covid-19 and energy access: An opportunity or a challenge for the African continent? *Energy Res. Soc. Sci.* **2020**, *68*, 1–4. [CrossRef]
80. Lying, K.; Brekke, A. Environmental life cycle assessment of biogas as a fuel for transport compared with alternative fuels. *Energies* **2019**, *12*, 532. [CrossRef]
81. Bušic, A.; Mardetko, N.; Kundas, S.; Morzak, G.; Belskaya, H.; Ivančičšantek, M.; Komes, D.; Novak, S.; Božidar, S. Bioethanol production from renewable raw material and its separation and purification: A review. *Food Technol. Biotechnol.* **2018**, *56*, 289–311. [CrossRef]
82. Sundvor, I.; López-Aparicio, S. Impact of bioethanol fuel implementation in transport based on modelled acetaldehyde concentration in the urban environment. *Sci. Total Environ.* **2014**, *496*, 100–106. [CrossRef] [PubMed]
83. Khuong, L.S.; Zulkifli, N.W.M.; Masjuki, H.H.; Niza Mohamad, E.; Arslan, A.; Mosarof, M.H.; Azham, A. A review on the effect of bioethanol dilution on the properties and performance of automotive lubricants in gasoline engines. *Royal Soc. Chem. Adv.* **2016**, *6*, 66847–66869. [CrossRef]
84. Kemausuor, F.; Adaramola, M.; Mórken, J. A review of commercial biogas systems and lessons for Africa. *Energies* **2018**, *11*, 2984. [CrossRef]

sustainability

MDPI

Article

Direct Improvement in the Combustion Chamber and the Radiant Surface to Reduce the Emission of Particles in Biomass Cooking Stoves Used in Araucanía, Chile

Robinson Betancourt Astete [1,2], Nicolás Gutiérrez-Cáceres [1,2,*], Marcela Muñoz-Catalán [1,2] and Tomas Mora-Chandia [1,2]

1 Mechanical Engineering Department, University of La Frontera, Temuco 4780000, Chile; robinson.betancourt@ufrontera.cl (R.B.A.); marcelanycoll.munoz@ufrontera.cl (M.M.-C.); tomas.mora@ufrontera.cl (T.M.-C.)
2 Center of Waste Management and Bioenergy, University of La Frontera, Temuco 4780000, Chile
* Correspondence: nicolas.gutierrez@ufrontera.cl; Tel.: +569-9857-45787

Abstract: Solid particle emissions from burning wood in three internal combustion biomass cooking stoves commonly used in southern Chile were compared. Each stove was used to show differences in sealing systems, combustion chamber shape, and heating surfaces in order to optimize biomass combustion and the energy produced at a low manufacturing cost. The influence of cooking stove design along with particle and gas emissions that resulted from the biomass combustion within the cooking stove was investigated in this study. Levels of diverse atmospheric contaminants, such as particulate matter, emission factor, NO_x, CO_2, and CO, and the temperature of the flue gases were determined with the Ch-28 method and UNE-EN 12815. The average emission of particulate matter was significantly reduced by modifying the geometry of the combustion chamber and heating surface of each stove, resulting in 5 g/h particle emissions in conventional equipment and 2 g/h in the improved equipment. In relation to gas emissions, there was a 25% maximum decrease in NO_x gases and 35% in CO after modifying the heating surface of each stove. This background supports the evidence of technological improvement with high environmental impact and low economic cost for local manufacturers.

Keywords: particle emission; biomass combustion; biomass cooking stoves; domestic heating

Citation: Betancourt Astete, R.; Gutiérrez-Cáceres, N.; Muñoz-Catalán, M.; Mora-Chandia, T. Direct Improvement in the Combustion Chamber and the Radiant Surface to Reduce the Emission of Particles in Biomass Cooking Stoves Used in Araucanía, Chile. *Sustainability* **2021**, *13*, 7205. https://doi.org/10.3390/su13137205

Academic Editors: Luis M. López-Ochoa and José P. Paredes-Sánchez

Received: 19 May 2021
Accepted: 15 June 2021
Published: 27 June 2021

1. Introduction

Chile is a country that is highly dependent on importing energy, particularly fossil hydrocarbons, even though it possesses a variety of energy resources that are relatively well distributed. Around 24% of the country's power grid comes from forest-based biomass [1], with firewood being the most-used energy source, mainly for heating and cooking purposes: 97% of firewood is used for heating, and the other 3% is used for domestic water heating and, in some cases, for cooking food [2]. The average annual consumption of firewood in a Chilean household depends greatly on the location, due to two fundamental aspects. The first is the geography of the country, where the rainiest and coldest areas are located mainly in the south, starting from the O'Higgins region to the Magallanes region, while the warmest areas are located in the north, starting from the Metropolitan region to the region of Arica y Parinacota. From west to east is the coastal area to the mountain range, where the mountain range zone sees more severe weather than the coastal area, which implies a higher consumption of firewood, encouraged by the abundance of biomass compared to other energy sources such as electricity, fossil fuels, and other sources such as geothermal waters, solar panels, and others. The second is the economic factor, since the country has large socio-economic differences that directly affect access to technologies and fuel for people in the south because the poverty rate is close to 17%, mainly in the Araucanía region [3]. This poverty rate in the area means that new technologies focused on house improvements, such as thermal insulation and efficient domestic heating, are not feasible for this percentage of the population. According to the latest government reports, fuel poverty in the region has reached 23% and 29% corresponding to the inhabitants without access to electricity supply and domestic hot water [4], which is directly proportional to the socio-economic status of each region; therefore, the few options for the most vulnerable part of the population to acquire biomass stoves are wood-burning stoves, due to its easy installation, versatility, and low price, which is a feasible alternative for this socio-economic sector.

At the national level, wood consumption, depending on the tree species, is obtained mainly from Hualle at 29%, which is equivalent to 3,435,890 m^3 st/year, followed by the Eucalyptus globulus with 24%, which is equivalent to 2,872,779 m^3 st/year, and finally the remainder corresponds to native and non-native species [5]. In the urban zones of the Aysén region, households consume an average of 18.3 m^3 of firewood a year. This goes down to 14.1 m^3 in Valdivia, La Union, Paillaco, and Rio Bueno (Los Rios region), 7.7 m^3 in Temuco and Padre Las Casas (La Araucanía region), and less in Chillan (Ñuble region) and Rancagua (O'Higgins region) [5]. One cubic meter of firewood is equivalent to approximately 700–900 kg, depending on the type of wood and its water content. Given that firewood is used by thousands of people during the year, wood burning has had severe social and environmental consequences in densely populated cities such as Temuco [2]. These consequences are mainly due to biomass combustion, which is an important source of particulate matter stemming from the incomplete combustion of components like cellulose, hemicellulose, and lignin, in addition to temperature-produced changes caused by combustion from uncontrolled sources [6,7]. According to air quality monitors [8], saturation conditions are present in more than a dozen cities in southern Chile. Saturation conditions occur when the maximum permitted concentration to which the population may be exposed to the environment has been exceeded. In Chile, the Ministry of Health has established air quality standards for coarse particulate matter (PM_{10}), D.S. N° 45/2001, and fine particulate matter ($PM_{2.5}$), D.S. N° 12/2011, where the maximum concentration allowed for $PM_{2.5}$ and PM_{10} per year is 20 and 50 $\mu g/m^3$ respectively. The permitted per day concentrations of exposure to PM should be lower than 50 $\mu g/m^3$ on average. Even in the face of the negative implications of firewood, its low cost in comparison to other fuels and the cultural tradition of its use in the country's southern cities make its replacement more difficult for the inhabitants. Implementing measures put in place by ruling governments, such as the atmospheric decontamination plan, the country has attempted to regulate the use of firewood and help create technological initiatives that directly impact the mitigation of contaminants by increasing the efficient and sustainable use of firewood fuels and prioritizing a reduction in atmospheric pollution, such as benefit programs for the replacement of cooking wood stoves, thermal conditioning improvements for housing [9], and to date, the PDTA-100857 project that provides the design and manufacturing so companies can continue improving their cooking wood stoves, on which this study is based.

The objective was to find the best design to improve biomass combustion in common wood-burning stoves in Chile, so that these design enhancements do not increase the manufacturing cost which can affect the final price of the product. Therefore, this design included a combustion chamber and heating surface. We demonstrate how these parameters influence the emission of atmospheric contaminants to help develop efficient, emission-reducing technology, increasing efficiency by 8%.

1.1. Biomass as a Fuel

Biomass is defined by the European Standardization Committee as a combination of organic matter derived from vegetable or animal sources or from their natural or artificial transformation, which may undergo energy treatments [10]. Specifically, solid forest biomass of lignocellulose origin is a product of natural and anthropogenic processes. The natural process involves the formation and growth in natural, water-based environments through photosynthesis, while the artificial process relates to formation through technological production and alterations to the previous natural constituents [11]. There are different types and forms of forest biomass that can be used as energy. On a global scale, 68% of the total bioenergy produced comes from forest biomass [12]. Firewood is cut into logs, ready to be used in domestic fuel apparatuses like stoves. Traditionally, firewood is used in households for cooking and heating. In general, the size of the logs is between 5 and 100 cm [13]. In Chile, the use of forest and agriculture biomass to generate electricity and thermal energy has varied over the last 12 years. On average, it represents 2.6% of the total energy produced [10]. It reached a high of 20% in 2011 and increased by 3% in 2016, coming in third place behind crude oil and carbon. As for firewood and its derivatives, they experienced a 14.7% increase in 2011 over the previous year, while in 2012 it increased 63% (60% residential use and 40% industrial fuel). The use of firewood is distributed between 82.1% in rural and 26.2% in urban homes [10]. Firewood is estimated to be consumed in 1,721,032 homes in Chile, equivalent to 11,926,411 m^3_{st} annually. If the national average of household energy is 10,232 kWh/year, including all fuels and electricity, then firewood represents 46.6% of the fuel used, or 4768 kWh/year. Almost all of its use is dedicated to heating [14]. This, together with the large development of the forestry industry, shows that firewood biomass is an important energy product with good future projections, with increasing demand and comparative advantages over other types of fuel (Figure 1).

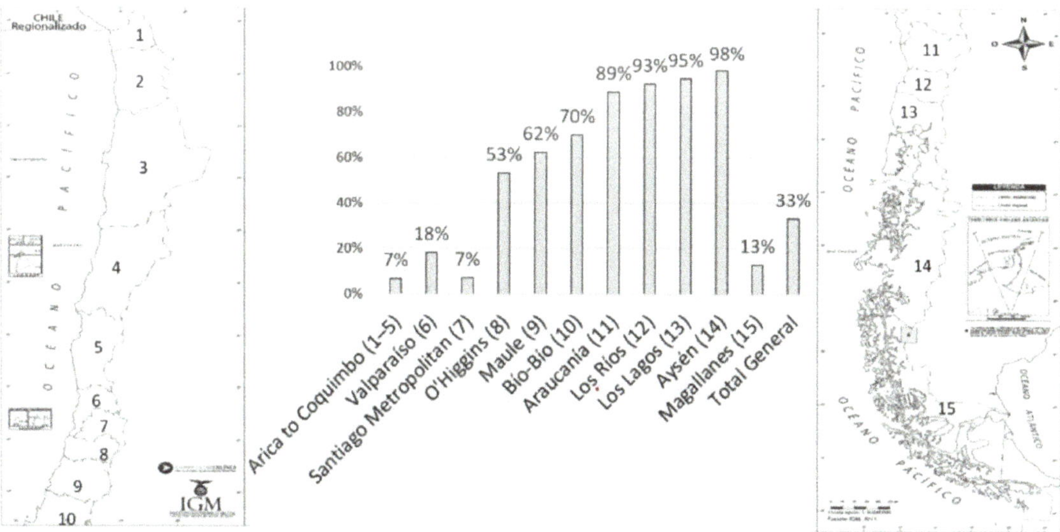

Figure 1. Household use of forest biomass in Chile [14].

Figure 1 shows the percentage of firewood consumption by region. Considering the availability of forest resources on a national scale, the highest percentage of firewood consumption and use is in the central-southern zone of the country. This is not surprising, as these are the climatologically colder zones and are located closest to the forest biomass production zones. According to the figures, consumption increases in the fall–winter months (May–August) and is mainly concentrated in the lowest socio-economic sectors [5].

The main wood-burning devices are salamander stoves, open chimneys, simple stoves, double chamber stoves, braziers, and handmade equipment. All of these devices are used, although some are more characteristic in certain areas. The wood fuel in these devices generates fine particulate matter emissions ($PM_{2.5}$), carbon monoxide, volatile components, nitrogen oxides, and other pollutants. As for CO_2 emissions, firewood is considered neutral [11].

Of the contaminants mentioned [8], the main problem in Chile is particulate matter ($PM_{2.5}$ and 86 PM_{10}), given the large impact on human health [15] that continuous exposure to these particles represents.

1.2. Biomass Combustion Process

Combustion of the wood starts when the biomass is exposed to caloric energy. It is followed sequentially by hydrolyzation, oxidation, drying, and pyrolysis, which increases the temperature to form a gaseous fuel. These substances are highly reactive and derived from carbon [16]. The process is as follows: 1. when the wood is exposed to a heating source, its elements start to hydrolyze, dehydrate, and burn as the temperature increases. This process produces volatile fuels, tarry substances, and highly reactive carbonaceous char. 2. When the ignition temperature of these volatiles and chemically treated substances is reached, the combustion process begins. 3. The heat generated by the combustion flame provides the necessary energy for the biomass to gasify and for the flame to spread, further evaporating the water that is found within the cell walls of the biomass, known as the water capillary action. 4. Then, the volatile products (such as water vapor, resinous compounds, and decomposing cell products), the hemicelluloses, and the lignin are separated to then be partially or completely combusted in the flame zone. During combustion, carbon continues to form until the flow of biomass gas falls below the minimum level required to keep the flames. During the flaming combustion, the formation of carbon continues until the volatile fuel flow drops below the minimum level required for the dispersion of the flame. 5. Finally, the smoldering process or the progressive oxidation of the reactive carbon starts. The biomass combustion is characterized by a non-premixed and turbulent flame, given that there is no mixing prior to the combustion reaction between the air and the fuel. In this process, large quantities of particles, which vary in size according to their physical and chemical properties, are emitted into the atmosphere. They can be divided into two categories: non-carbon-based particles, which are generated by non-flammable elements within the fuel, possess neither carbon nor hydrogen atoms [17], and include residual elements that detach from the surface of the fuel [18]. The second category consists of carbon particles formed by pyrolysis of the fuel molecules, and which have not reacted in the flame zone. Other conditions that contribute to the production of solid particles mainly depend on the level of humidity found in the fuel. At some point, moisture in biomass residues significantly affects your ability to heat [19].

2. Materials and Methods

This paper investigated three different types of cooking wood stoves commonly used in Chile, corresponding to cooking stove A, B, and C, the descriptions for which are mentioned below. The construction materials are defined in Table 1.

Table 1. Construction materials in each device.

Biomass Cooking Stove	Combustion Chamber	Sealed	Radiant Surface
A	Cement	Not sealed	Gray cast iron
B	Cement	Glass rope lagging	Gray cast iron
C	Cement	Glass rope lagging	Gray cast iron

Stove A is a conventional model, which can be found in the national market. Stove B is similar to the previous model, except that it has a different cover that allows a hermetic seal, making it more efficient than stove A. The C stove is similar to the B stove except for the combustion chamber, which is delimited by a fin as shown in Figure 2, which reduces the area of the flue gases by 60% over its previous versions. It is considered that only stove A is found for sale in the national market, while the other two are improved prototypes derived from the A device; therefore, those devices are not available in the domestic market yet.

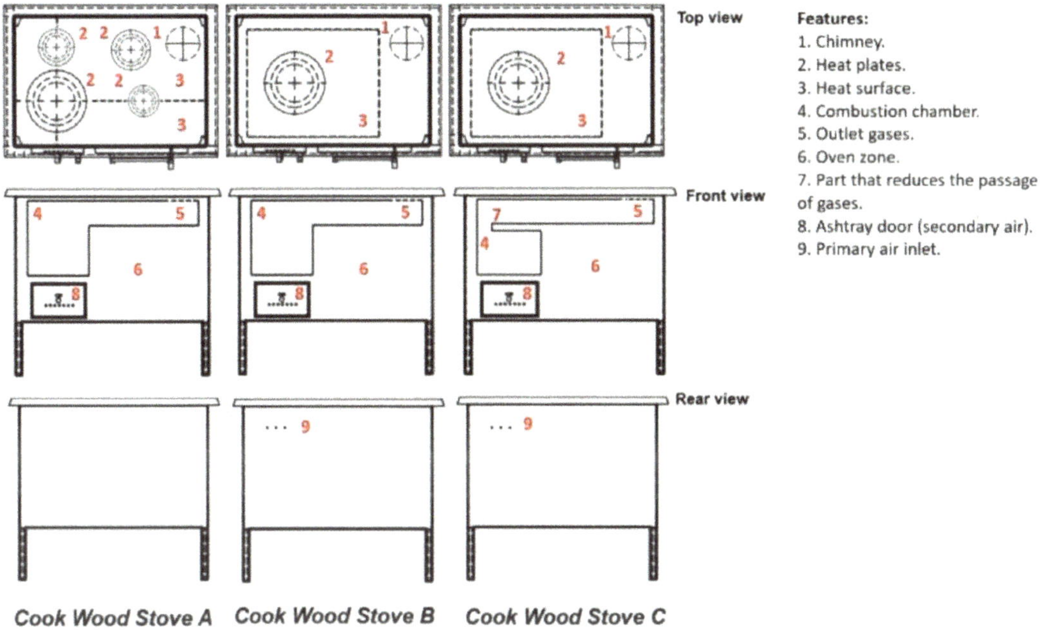

Figure 2. Designs of each device used to evaluate the combustion process.

The parameters of the heating surface and combustion chambers of each device are presented in Figure 2. In the case of stove A, the heating surface is divided into six parts: two bigger parts that hold the surface together, and four smaller, circular shapes (heat plates). In stoves B and C, the surfaces are the same, with the surface molded into three bigger parts that secure the surface together with a smaller circular-shaped surface.

The manufacturer has not published the thermal power or efficiency of the devices; hence, one of the main objectives of this research was to determine their performance.

2.1. Properties and Improvement in Each Model

The improved stove designs (B and C) were compared to the original cooking stove design A, so the differences between each device are divided into three aspects: air inlets, sealing of the device, and combustion chamber geometry.

2.1.1. Air Inlet

The air inlets for devices A, B and C are presented with a yellow arrow, while the red arrow represents parasitic air inlet, in Figure 3.

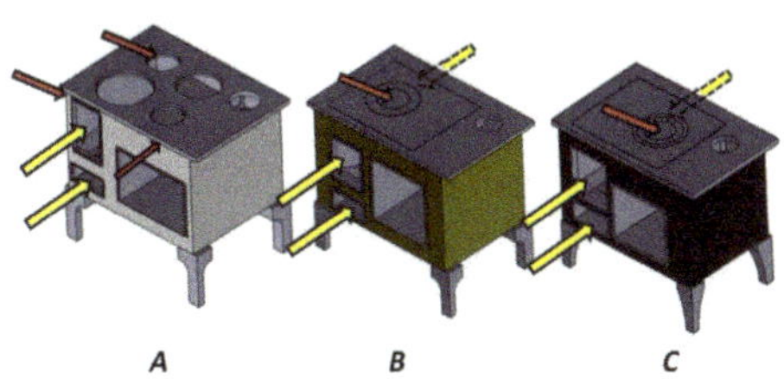

Figure 3. The air inlet are pointed by yellow arrows, while the parasite air inlets and unwanted gas outlet are pointed by red arrows, in each stove.

Stove A initially presents two air inlets: the first in the fuel load door and the second in ashtray door. However, the manner of craft manufacture that predominates in these types of devices, which lacks exhaustive reviews and quality inspections, affects the prevalence of manufacturing errors such as fissures between material joints and bindings. As an example, the space between the combustion chamber and the radiant surface allows the parasite air to get in and the unintended combustion gases to get out. Stoves B and C, in this respect, were built in identical ways and have two air inlets; the former has seven holes in the ashtray door, which can regulate its opening through a lever device, while the latter has three holes located in the rear side of the device at the height of the combustion chamber. It is assumed that devices B and C also have parasite air inlets and unwanted gas outlets that were not evaluated in this investigation, shown with red arrows in Figure 3. The air inlets in the ashtray door feed the combustion process in order to keep the fuel burning. The problem is that when the embers and ashes obstruct the air flow on the grate, the combustion process is affected due to the lack of air. To avoid this, three air inlets were added in the rear part of the device (see Figure 2, Feature 9), which maintains the minimal conditions of combustion.

The air flow and movement of combustion gases are seen in Figure 4 and the scheme is the same in each stove, where the passage of the combustion gases through the output duct is regulated by a damper that is entirely closed in all the tests. Consequently, the combustion gases move around the presented section.

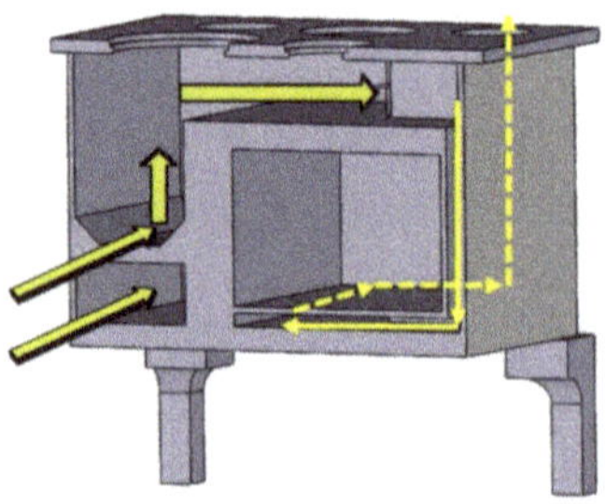

Figure 4. Diagram of the flow of air and gases in each stove.

2.1.2. Sealing

The optimal sealing for each device is according to the parasite air inlets and the unwanted gas outlets. Considering this, the material used to seal those fissures is fiberglass cord, which can withstand high temperatures without reacting to the flame. Fiberglass cord was used to improve the sealing in the fuel load door and the ashtray door of stoves B and C while the A stove does not have this form of sealing. The sealing also considers the new geometry of the radiant surface: the dilatation and the inherent weight of that surface are enough to create a labyrinth seal to avoid the inlet and outlet of air. The surface's joint type is shown in Figure 5, where the moving part corresponds to the hot plates and the radiant surface (see Figure 2, Features 2 and 3, respectively).

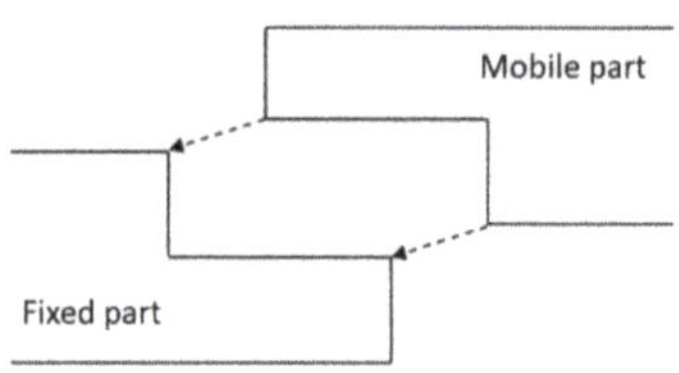

Figure 5. Superficial sealing in cooking wood stove B and C.

2.1.3. Combustion Chamber Geometry

The chamber combustion design has been limited due to the technical feasibility and the manufacturing that most of the manufacturers in the Araucanía region have. Otherwise, the design changes have been applied only to the C device, whose scheme and differences are presented in Figure 6.

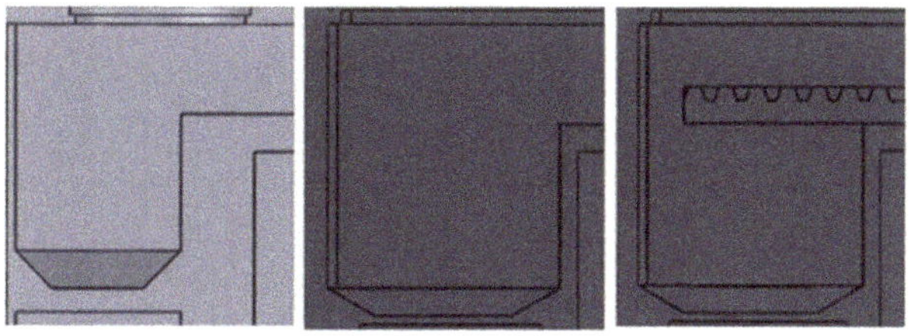

Figure 6. Combustion chamber geometry in each stove.

The available combustion chamber volumes in each stove are 0.0343 m^3 for stoves A and B, and 0.0214 m^3 for stove C, which is much smaller due to the closure made in the combustion chamber that permits a smaller amount of test fuel. From the point of view of dimensions, cooking stoves A and B have similar dimensions in height, width, and length, while cooking stove C has a lower height due to the piece added on the surface of the combustion chamber, reducing its height by 8 cm. The combustion chamber's changes not only consider the geometry and design of the walls, but also the grate from which the ashes descend. Stoves A, B, and C have a conventional grate design. Those grates are presented in Figure 7.

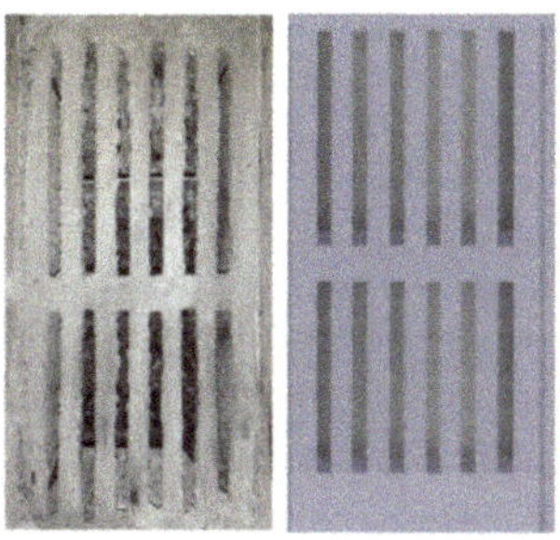

Figure 7. Grate geometry, where the ashes descend in each stove.

2.2. Properties of Biomass Used

The biomass used was Eucalyptus globulus, which was treated to maintain homogeneity in dimensions, mass, and similar moisture content. Its properties are listed in Table 2. The Chilean method used for this was 5G [20].

Table 2. Description of samples.

Biomass Cooking Stove	Fuel Type	Physical Description
A	Biomass/sample test wood treated	310 × 50.8 × 101.6 mm
B	Biomass/sample test wood treated	300 × 50.8 × 100.0 mm
C	Biomass/sample test wood treated	305 × 50.8 × 101.0 mm

The fuel used in the study was chemically analyzed using two forms of analysis: the first consisted of an elemental and thermogravimetric analysis (Table 3); the second was the fuel load and moisture test (Table 4). The results will be shared next. The chemical properties are expressed in the next table for each sample.

Table 3. Thermogravimetric and elemental analysis of Eucalyptus globulus.

Thermogravimetry of *Eucalyptus globulus*					
Volatile (%)		Fixed Carbon (%)		Ashes (%)	
81.6		17.1		1.1	
Elemental analysis of *Eucalyptus globulus*					
C (%)	H (%)	O (%)	N (%)	S (%)	Cl (%)
48.1	5.9	44.1	0.3	0.0	0.2

Table 4. Fuel load and moisture (% dry basis) in the test sample.

Sample	Moisture Content (%)	Std. Dev.	Fuel Load (kg)	Std. Dev.
Biomass for A	13.9	0.9	2.7	1.3
Biomass for B	15.0	0.9	2.9	0.3
Biomass for C	15.8	1.3	2.2	0.8

2.3. Test Method

The experiments were carried out randomly with a previous sampling frame. The methodology used to extract samples from the combustion particulate matter was from the 5G method, and the variables that were set for the use of the cooking wood stove were from studies that mention the predominance of their use, such as the secondary air inlet, which was tested in a single position (lower potential), closed air inlet [21]. The sample was extracted from the chimney or gas evacuation ducts, while the particle extraction was from the dilution duct, given that the hood captures all emissions. Figure 8 shows this process. To extract and sample the combustion gases, a Testo 350 gas analyzer was used, the methodology of which can be found in the NCh3173 and UNE-EN 12815 norms [22]. As for the particle extraction, this was done by collecting samples in 110 mm diameter fiberglass filters made by Pall Corporation. An isokinetic methodology, 5G, was used by the Environmental Supply team to extract the particles. Each stove was used at its lowest potential and tests were repeated six times for each device, giving a total of 12 filters per device and 36 filters across all of the tests. The filters were collected at the end of the total combustion of each device's fuel.

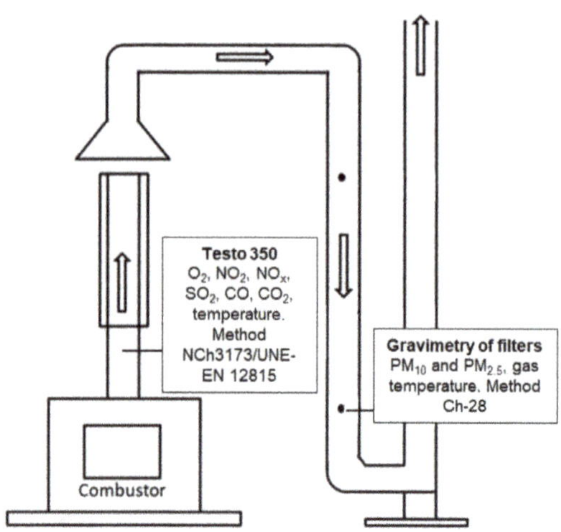

Figure 8. Diagram and instrumental design of the gas and particle measurement system.

In relation to the methodology acquisition and monitoring temperature, four Thermocouples Type K temperature sensors of ceramic fiber were used, of which acquisition and monitoring was through the CompactDAQ 9213 and Labview 2011 hardware and software, both from the National Instruments Company. It is worth mentioning that thermocouples are not exposed to direct flame; therefore, correction for radiative temperature is not required. The sampling time was equivalent to a measurement for 1 min, whereas the position of the sensors was the same for each device, located in the radiating surface, combustion chamber, the flame region/area, and in the oven of the different devices, as shown in Figure 9.

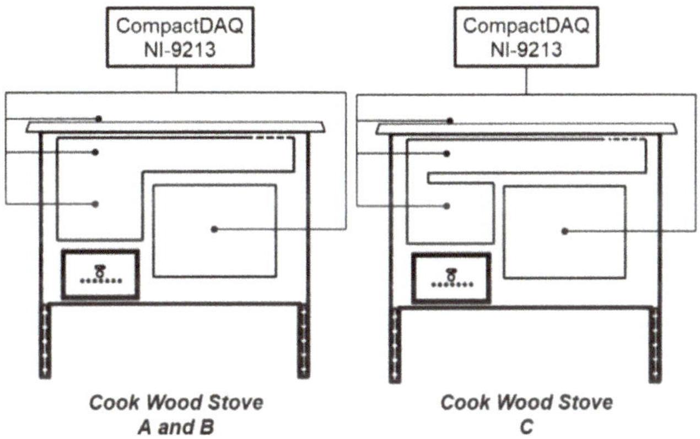

Figure 9. Diagram and instrumental design of acquisition and temperature measurement system.

2.3.1. Particulate Matter

The calculations of the particulate matter emissions for each device were solved in detail according to the Chilean 5G and Ch-28 methodology and were obtained by the following equation.

$$E = C_s \cdot Q_{std} \tag{1}$$

where E corresponds to the particulate matter emission rate. C_S and Q_{std} are the equivalent to the particulate matter concentrations in the flue gas on a dry basis, which were rectified under standard conditions, and to the average flow rate of the gas flow in the dilution tunnel, respectively. Both values were already calculated using the indicated methods, and it was not the objective of this investigation to explain how the values were obtained. Finally, the emission rate should be applied in the adjustment factor according to the following equation:

$$E_{adj} = 1.82 \cdot E^{0.83} \tag{2}$$

The adjustment factor emission is based on a statistical adjustment that comes from the Monte Carlo simulation defined for population ranges or sample quantities. The details of this factor are presented in the report [23].

2.3.2. Performance Measurement

The results of the different combustion processes, referring to the performance evaluation that was carried out for each device, were obtained experimentally through test runs. The combustion efficiency requires the sensitive and the latent heat losses in fumes, in addition to the unburned fuel losses, in which each previously mentioned variable is expressed hereafter.

$$Q_a = (t_a - t_r) \cdot \left[\frac{C_{pmd} \cdot (C - C_r)}{0.536 \cdot (CO + CO_2)} + \left(C_{pmH_2O} \cdot 1.92 \cdot \frac{(9H + W)}{100} \right) \right] \tag{3}$$

where t_a and t_r are equivalent to the combustion gases and ambient temperature, respectively. C_{pmd} corresponds to the specific heat of dry combustion gases in standard conditions, based on the temperature and gas composition. In addition to that, C_{pmH_2O} corresponds to the water-specific heat in standard conditions depending on the temperature. Moreover, CO and CO_2 are the carbon monoxide content and the carbon dioxide content in dry combustion gases, respectively. At last, C, H, and W are the carbon, hydrogen, and the moisture contents in the test fuel.

$$Q_b = 12644 \cdot \left[\frac{(C - C_r)}{0.536 \cdot 100 \cdot (CO + CO_2)} \right] \tag{4}$$

C_r is equal to the carbon content contained in the residues, regarding the amount of fuel burnt, the approximation for which is given by $C_r = R \cdot b/100$. In addition, b and R correspond to the fuel components in the residues in relation to the residue material mass and the residue that goes through the grate with regard to the test fuel. In this study, it was measured in each sample.

$$Q_r = 335 \cdot \left[\frac{R}{100} \right] \tag{5}$$

The combustion efficiency was calculated as:

$$\eta = 100 - \left(\frac{100}{H_u} \cdot (Q_a + Q_b + Q_r) \right) \tag{6}$$

To determine the total thermal power of each device according to the same methodology, the following equation was used:

$$P = \left(\frac{H_u \cdot B \cdot \eta}{100 \cdot 3600} \right) \tag{7}$$

where B is the amount of test mass and H_u is the lower calorific value of the test fuel.

2.3.3. Emission Factor

The emission factor represents the number of pollutants emitted into the atmosphere, associated with the activity that generates pollutants [24]. Thus, the emission factor for each device, considering the average emission of particulate matter per fuel quantity, was used for this study, as shown in the following equation.

$$EF = \frac{E_{adj}}{B} \tag{8}$$

3. Results

3.1. Main Emissions from Combustion

The main emissions of the combustion process correspond to CO, NO_x, and PM_{total}, which are essential to be able to evaluate the thermal behavior of the appliance and to obtain the respective emission factor or EF_{PM} for each cooking stove, expressed in the summary in Table 5. The reference O_2 was 13% in the gas measurement for all samples, and the uncertainty of the gas measurement was less than 5%. The differences of units between concentrations of emission gases and particulate matter were defined by the methods used.

Table 5. Summary of emissions and emission factor for each stove.

	Cooking Stove A			Cooking Stove B			Cooking Stove C		
	Sum. Avg.	**Std. Dev.**	**U**	**Sum. Avg.**	**Std. Dev.**	**U**	**Sum. Avg.**	**Std. Dev.**	**U**
CO (mg/Nm^3)	30.3	10.3	±2.0	34.3	16.6	±2.0	19.7	7.6	±2.0
NO_x (mg/Nm^3)	4999.2	959	±10	4456.2	1293.2	±10	3748.2	260.8	±10
PM_{total} (g/h)	5.4	2.7	±0.6	2.1	0.6	±0.3	4.4	1.8	±0.24
EF_{PM} (g/kg)	2.7	1.3	-	0.9	0.2	-	2.0	0.8	-
	Cooking Stove A			**Cooking Stove B**			**Cooking Stove C**		
	Avg.	**Std. Dev.**	**U**	**Avg.**	**Std. Dev.**	**U**	**Avg.**	**Std. Dev.**	**U**
CO (mg/Nm^3)	6.081	2.115	±2.0	6.008	3.681	±2.0	5.125	3.208	±2.0
NO_x (mg/Nm^3)	87	35	±10	63	10	±10	69	25	±10
PM_{total} (g/h)	5.4	2.7	±0.6	2.1	0.6	±0.3	4.4	1.8	±0.24
EF_{PM} (g/kg)	2.7	1.3	-	0.9	0.2	-	2.0	0.8	-

The CO emissions caused by biomass combustion can be used to indicate the amount of oxygen present in the reaction process. As the fuel evaporates and its mass falls, CO emissions are not reduced. This is because the biomass stove allows air to enter the combustion chamber in uncontrolled proportions, avoiding reaching a thermodynamic state in equilibrium; this situation is apparent in the high variability of emission values in A, and to a lesser extent in B. This situation is reflected in Figure 10, which represents the range of concentration variability emitted by each stove, where C is the one with the lowest emissive trend.

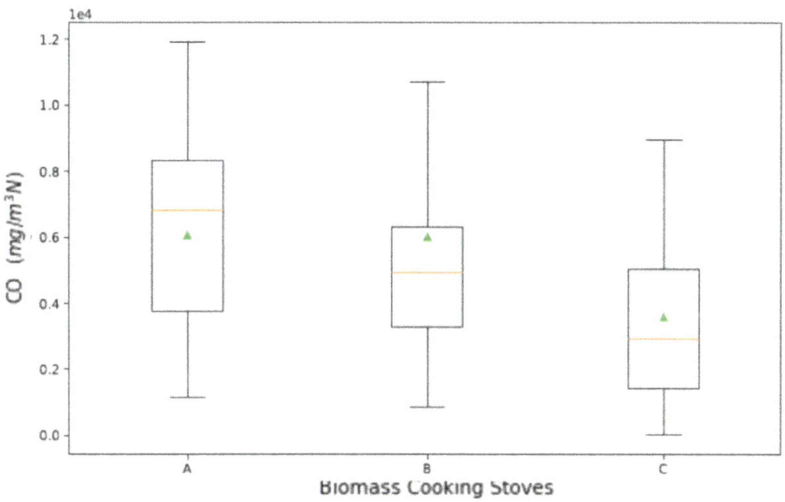

Figure 10. Box plot representation of the average CO emission of each stove.

Concentrations emitted per hour of average operation for each cooking stove were 6081, 6008, and 5125 (mg/Nm^3), respectively. In comparison with other studies with wood cooking stoves, Koyuncu et al. [25] reported that the emissions of a domestic heating system with similar features were 1489 (mg/MJ), and studies carried out by Boman et al. [26] reported emissions for a different variety of fuel between the ranges of 580 to 2340 (mg/MJ). In this investigation, they were of 1254, 1205, and 1028 (mg/MJ) for CO, the difference of which is between 15 and 30% of what Koyuncu et al. [25] mentioned. This shows evidence of a significant reduction of these emissions. The relationship between the average concentration of CO emitted and the excess air of the device in each test is also given (Figure 11).

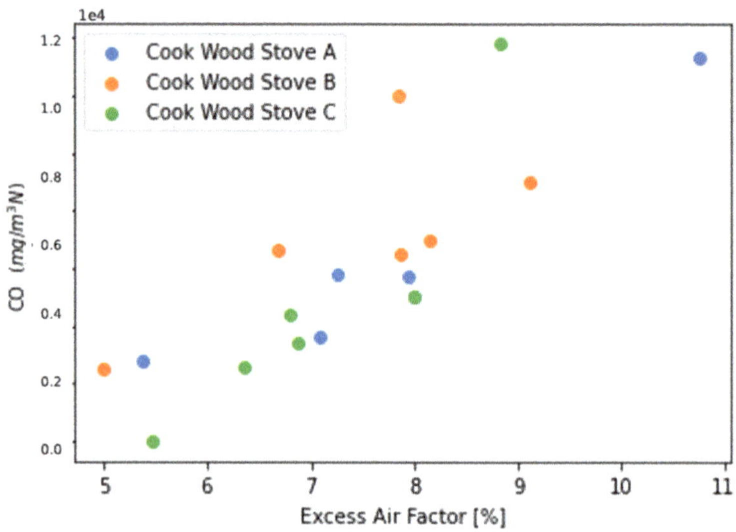

Figure 11. Relationship between the average concentration of CO emitted and the excess air measured in each stove.

In Figure 11, the greatest number of points close to each other was obtained with the type C cooking wood stove (green points), while there was a clear dispersion in the results of A and B, meaning that device C exhibited a stable behavior and more efficient combustion due to the relationship between the emission of CO versus the excess air, which means that it used less air for the combustion process and allowed the reduction of emissions of CO. Therefore, it is shown that the emission trend of CO is strongly linked to the air/gas leaks present in the combustion chamber. In summary, the emission behavior from the highest to the lowest was demonstrated by device C, followed by B, and finally by A. Despite the fact that cooking stoves B and C are similar, the throttling at the outlet of the combustion chamber in C (Figure 3) allows combustion to be carried out with less air, which demonstrates the low CO emissions, since the combustion process carried out is more efficient.

The possible gas phase reaction mechanisms for NO_x formation in combustion is described by three mechanisms [27–29]: 1. thermal NO_x is caused by temperatures over 1800 K, which reacts with atmospheric N_2 in the combustion chamber; 2. fast NO_x created in the front of the flame; and 3. NO_x is caused by the N_2 content in the fuel. In the case of the combustion addressed by this study, it only interacts with the NO_x of the fuel [27,28]. Thus, NO_x emissions obtained in the measurement process represent the formation and reaction of N_2 present in the fuel, with a linear correlation between the oxidized N_2 and the NO_x emissions [24]. A summary of the NO_x emissions for each cooking stove (Figure 12) is presented.

Figure 12. Dispersion of the average NO_x emission values in each stove.

The concentrations emitted per hour of average operation of NO_x between stoves A, B, and C were 4999.2, 4456.2, and 3748.2 (mg/Nm^3) per hour of sample, respectively. Therefore, a difference or decrease in the emission of NO_x can be ensured. In relation to similar combustion systems, and following the reports presented by Koyuncu et al. [25] where NO_x was 12.54 (mg/MJ), and the results in this case were 10.8, 11.4, and 10.2 (mg/MJ), which differ by 13%, 8%, and 18%, respectively.

As stated previously, the NO_x emissions are associated with the N_2 content present in the fuel; however, the differences presented in the assays of cooking stove A in contrast to B and C were substantial. This is due to the non-controlled combustion process itself shown in A, since it has a large number of air inlets, making a stable combustion over time more difficult. This implies the presence of high temperature peaks that can occur inside the flame, causing the formation of thermal NO_x, which depends both on the amount of O_2 and N_2 and on the fuel present in the reaction [30]. The low NO_x emission in C is attributable mainly to the throttle of the gas outlet inside the combustion chamber, which means that the combustion process occurred with a lower amount of air contributing to an efficient combustion.

The particulate matter emissions are also affected by design changes, both in the combustion chamber and in the heating surface. However, the influence of the moisture content of the fuel on the emission of PM is not significant when it is close to 14 ± 3.6% on a dry basis [31], i.e., there is no statistical support that a variation in humidity like that obtained in the samples significantly affects the emission of particles. A reduction in PM_{total} emission of 62% and 18% was obtained for device B and C, respectively, compared to cooking stove A, as shown in Table 5. Taking other investigations with a different result, Chen et al. [32] reported an EF_{PM10} of 18.1 ± 6.6 (g/kg) and 12.7 ± 1.26 for $EF_{PM2.5}$. On the other hand, Cooper et al. [33,34] reported that the $EF_{PMtotal}$ for biomass stoves was near to 8.5 (g/kg), demonstrating that the suggested improvements reduce the emission of total particulate matter. It should be considered that stove A does not have seals on its doors, and the cover it uses is, at various points, open to the sample environment, so the results obtained are not accurate, as an unquantified number of particles may be leaving the test environment. Despite the knowledge that this situation can occur, the test was performed in the same way to have a reference, because there is no methodology that combines the measurement of particles emitted by the stove through the chimney and into the testing environment. The results obtained by B and C presented a much smaller standard deviation than that obtained in A, because the methodology can be applied under the characteristics of B and C, i.e., the results of B and C were valid according to the methodology used. The emission factor was reduced by 66% and 25% for B and C, respectively.

The distribution of total particle emissions for each device is presented in Figure 13. It is evident that stove B was the one with the lowest emission rate compared to the others. This situation is motivated by the type of seal that it has both in the deck and on the door. Stove C, whose characteristics are similar, did not present a concentration lower than B, however, due to the throttling and the area through which the secondary air enters the combustion chamber. Moreover, there is the possibility of producing a second phase of gas oxidation and this generates the second pyrolysis reaction of the PAH, which forms PM_{Total}.

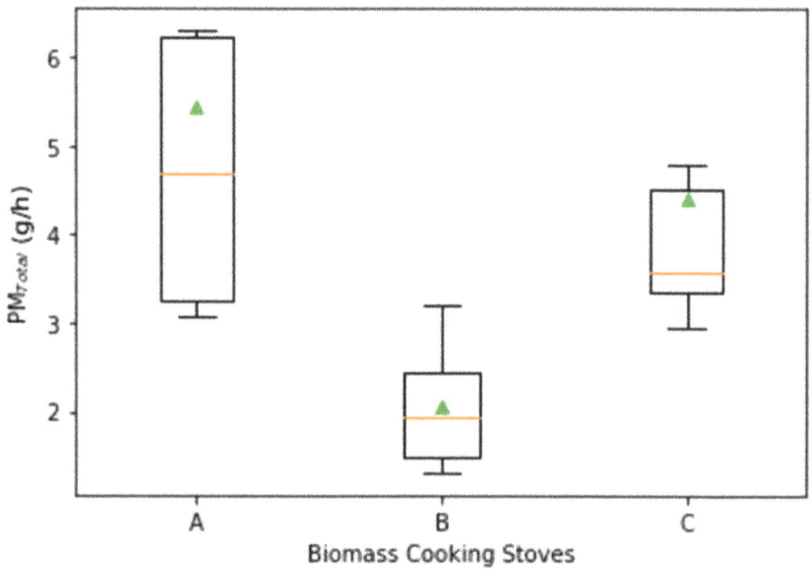

Figure 13. Total emission distribution of particulate matter in each stove.

3.2. *Thermal Behavior of Each Stove*

The results of the thermal behavior of the stoves vary depending on the type of surface they have. Devices like wood-burning stoves are designed to raise their surfaces' temperatures, mainly to enable cooking or heating of the surrounding area (see Table 6). The best thermal behavior, then, is found in the device that can maintain constantly high temperatures throughout the sample time. For this test, that device was cooking stove B, which reached a maximum temperature of 192.3 °C, followed by stove C, and, finally, stove A. The high temperatures in stove B can be explained by its combustion chamber, which does not include a system for enclosing flames, so the flames are allowed to pass directly to the stove surface. Similar results were reported by Hueglin et al. [35], in which the thermal behavior of devices that use wood as a fuel were mentioned. Stove C, on the other hand, has a closed combustion chamber that partially retains the flame, as seen in Figure 2. However, these results are not sufficient to define the efficiency of each device.

Table 6. Summary of the behavior of the temperature in each biomass cooking stove.

Cooking Stove	Surface (°C)			Combustion Chamber (°C)			Exhaust Gases (°C)		
	Avg.	Std. Dev.	Uncertainty	Avg.	Std. Dev.	Uncertainty	Avg.	Std. Dev.	Uncertainty
A	169.5	6.9	±8.4	420.0	75.1	±10.5	1690.5	24.0	±4.7
B	184.9	5.0	±9.2	446.5	47.8	±11.1	154.0	10.8	±3.8
C	177.5	5.8	±8.8	578.7	63.8	±14.4	144.6	14.4	±3.6

Combustion gas emissions depend on the combustion process carried out in each device. The resulting temperatures in the combustion chambers are shown in Figure 14. The temperatures in the combustion chambers were noticeably higher in stove C. This can be explained by the shape of the chamber that allows the flame area to have a higher temperature over time. This shows that by carrying out combustion processes in smaller areas and with stable air entrances, it is possible to obtain reactions in elements present in the gases released by the flame [36]. Another variable to determine the thermal behavior in each cooking stove is the temperature of the combustion gas or exhaust temperature, as this temperature is directly related to the emissions and depends on the combustion process carried out in each device. The resulting temperatures in the combustion gas are shown in Figure 14. Exhaust temperature can be an indicator of a stove's thermal efficiency. For stove A, the average temperature was 190 °C. The temperatures for stoves B and C were significantly lower (see Figure 14), with average values of 154 °C and 144 °C, respectively.

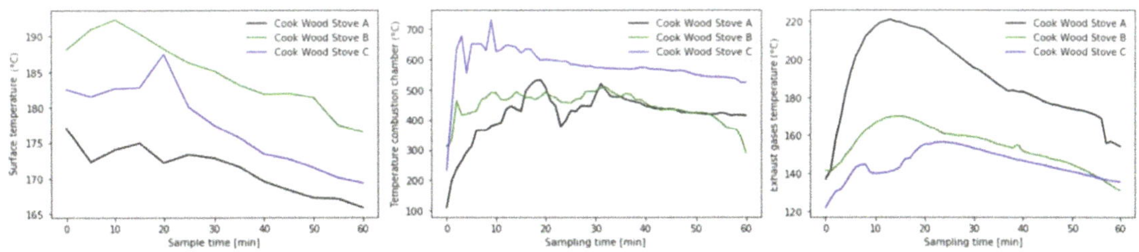

Figure 14. Temperature of the heating surface (**left**), combustion chambers (**center**), and exhaust gases temperatures (**right**) during the measurement process for the three stoves.

Studying the temperature of the exhaust gases can highlight two aspects: 1. the time in which the combustion fuels are kept within the combustion space, where reactions of solid and gaseous compounds that do not completely oxidize during the combustion process are favored, which releases energy by radiation to the device's combustion chamber; and 2. the type of heating surface that controls the flame in the stove. High temperatures in the gases and little oxygen can also cause the formation of solid elements. These elements can react to larger particles due to their soot cores [37]. In this study, only point 2 was considered; thus, the stove A showed higher temperatures than stoves B and C, demonstrating lower fuel efficiency, as shown in Table 7. The temperatures found in stoves B and C were similar due to the kind of cover that they have. The shape of stove C had less influence, 7%, on the temperature of the gases than stove B.

Table 7. Results of the combustion process in each stove.

Cooking Stove	Thermal Power (kW)	Std. Dev.	Burning Rate (kg/h)	Std. Dev.	Efficiency (%)	Std. Dev.
A	6.5	1.4	1.6	0.3	73.4	1.8
B	5.6	1.2	1.4	0.2	48.7	1.5
C	7.4	1.5	1.7	0.2	79.7	2.8

3.3. Performance of Stoves

The results on stove performance and behavior are expressed in Table 7.

The biomass combustion occurring in each stove was significantly different in terms of gas emissions, particulate matter, and also in the temperatures released from the heating surface. Furthermore, the average burning rates were also different for each one, showing a reduction of 12.5% in stove B and an increase of 6.3% in stove C. With the modification of the combustion chamber, the seals of the device and the geometry of the surface increased the efficiency of the devices B and C between 5% and 6%, respectively, as shown in Table 7. The performance, thermal power, and emission of particles of each tested device are expressed in the following graphs with an overview of the values in Figure 15.

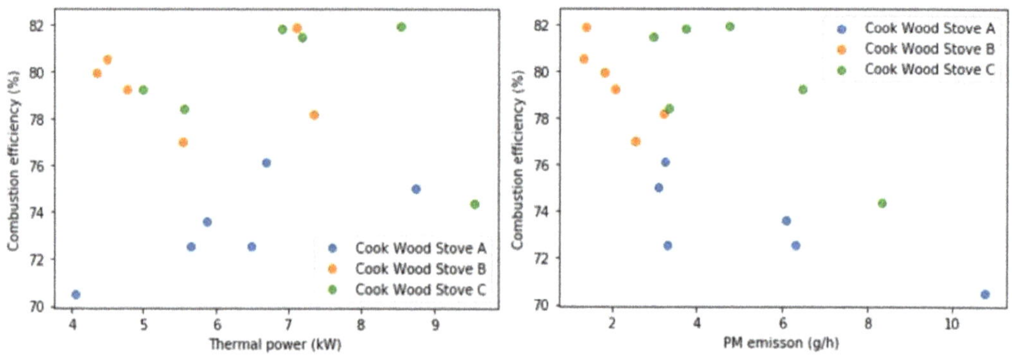

Figure 15. Chart combustion efficiency versus thermal power (**left**) and PM emission (**right**) in each stove in all samples.

The relation between combustion efficiency and the thermal power emitted shows that device A is the one that had the lowest relation between efficiency and power, while stove C had the best performance. Their cumulative performance in ascending order was equivalent to 73, 79, and 80% for stoves A, B, and C, respectively. The comparison between the combustion performance and the total cumulative emission of PM shows that stove B performed better than A and C. This comparison showed that higher thermal power and lower combustion performance produce a higher PM emission. The comparison between the combustion performance and the total cumulative PM emission shows that stove B and C performed better than A, which was expected, but through Figure 15, a more stable behavior could be established for stove B, in addition to presenting a smaller dispersion in its results of combustion efficiency versus PM emission. This comparison also showed that higher thermal power and lower combustion efficiency produce a higher PM emission. According to the thermal power generated by each stove and their relation to the burned material rates, stove C is the one that presented the best results since it presented the lowest data dispersion. Even though the power increased with the burning rate, this does not ensure that the process is effective, as shown in the graphs for stove A. The statistical analysis using a Student's t-test with a 10% confidence interval provided statistical significance between the samples, which shows that there are changes in the emission factors. However, stove C showed in the graph in Figure 16 that it possessed a lower emission factor to a specific power, whereas stove B behaved similarly to stove C. This was not the case for stove A, which showed greater dispersion data of the emission generated by the power produced.

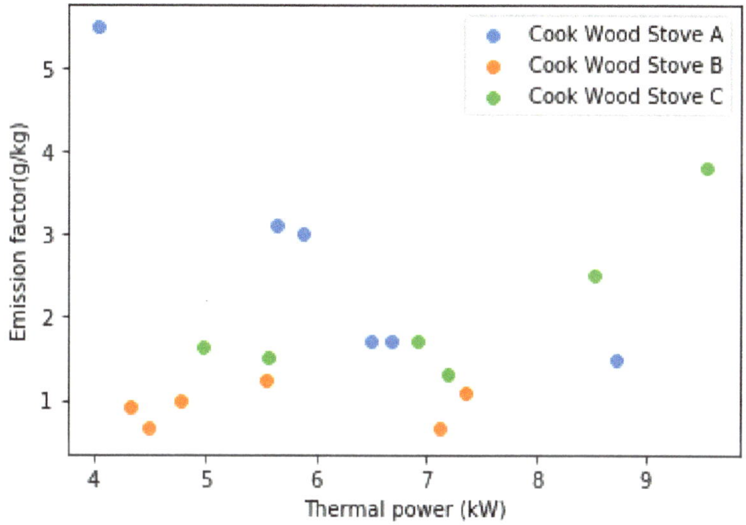

Figure 16. Chart emission factor versus thermal power in each stove in all samples.

4. Conclusions

The decrease in the formation of particles of versions B and C (compared to the traditional one) was achieved through combustion air control and the adjusted volume of the combustion chamber. This enabled a better use of fuel power, which permitted the reaction of a higher number of reactive volatiles present in the gases, as shown in Figures 11 and 13.

Modifying the heating surface showed that less air was filtered into the interior of the combustion area. This is due to the number of elements, from thirteen to five elements (Figure 2), which comprised the surface and guaranteed a better control of the biomass combustion process, represented in the efficiency obtained (see Table 7).

With the modifications presented in the manuscript, it is possible to reduce the total particulate matter (PM_{total}) emissions by 63%, followed by gas emissions, with a maximum reduction of NO_x and CO gases by 25% and 35%, respectively.

Finally, it is concluded that combustion chamber design and heating surfaces favor a biomass combustion process that is much more efficient than the conventional version. This improvement does not suppose an increase greater than 2% of the production cost. Therefore, this technological information has a positive impact on the region, since it is a feasible change that more small and medium-sized manufacturing local companies in Temuco can implement in their designs, which are available to citizens, the regional government, and manufacturers.

Author Contributions: In this research work, the contributions of each author are as follows: conceptualization by R.B.A.; methodology and experimentation by N.G.-C. and M.M.-C.; data validated and analyzed by T.M.-C.; research by R.B.A., N.G.-C., and M.M.-C.; R.B.A. contributed material tools and resources; manuscript writing by R.B.A., N.G.-C., and M.M.-C.; review and supervision by T.M.-C. and R.B.A. All authors have read and agreed to the published version of the manuscript.

Funding: This research received no external funding.

Institutional Review Board Statement: Not applicable.

Informed Consent Statement: Not applicable.

Data Availability Statement: The information and database for this research are currently not on a platform or website. They can be provided by the corresponding author.

Acknowledgments: This research was conducted thanks to the contributions of the Faculty of Engineering and Sciences of Universidad de La Frontera, Temuco, Chile, together with the Laboratory of Combustion and Particles of the Department of Mechanical Engineering and the Center of Waste Management and Bioenergy of the Universidad de La Frontera.

Conflicts of Interest: The funders had no role in the design of the study; in the collection, analyses, or interpretation of data; in the writing of the manuscript, or in the decision to publish the results.

References

1. Ministry of Energy—Chile. *National Energy Balance*; Technical Report; Ministry of Energy—Chile: Santiago, Chile, 2015; p. 412.
2. Ministry of Energy—Chile. *Policy of Firewood Use and His Derivatives for Heating*; Technical Report; Ministry of Energy—Chile: Santiago, Chile, 2015; p. 414.
3. Ministerio de Desarrollo Social y Familia. *Desarrollo Social 2019*; Technical Report; Gobierno de Chile: Santiago, Chile, 2019.
4. Amigo, C.; Araya, P.; Billi, M. *Políticas Públicas y Pobreza Energética en Chile: una Relación Fragmentada*, 1st ed.; RedPE: Santiago, Chile, 2018.
5. Chilean Construction Chamber. *Measurement of National Consumption of Firewood and Other Solid Fuels Derived from Wood*; Technical Report; Ministerio de Energía: Santiago, Chile, 2015.
6. Cereceda-Balic, F.; Fadic, X.; Llanos, A.L.; Domínguez, A.M.; Guevara, J.L.; Vidal, V.; Díaz-Robles, L.A.; Schiappacasse, L.N.; Etcharren, P. Obtaining polycyclic aromatic hydrocarbon concentration ratios and molecular markers for residential wood combustion: Temuco, a case study. *J. Air Waste Manag. Assoc.* **2012**, *62*, 44–51. [CrossRef] [PubMed]
7. Medeiros, P.M.; Simoneit, B.R. Source profiles of organic compounds emitted upon combustion of green vegetation from temperate climate forests. *Environ. Sci. Technol.* **2008**, *42*, 8310–8316. [CrossRef] [PubMed]
8. Ministry of Environmental—Chile. *Atmospheric Decontamination Plan (PDA). Strategy 2014–2018*; Technical Report; Ministry of the Environmental Chile: Santiago, Chile, 2014.
9. Resolution 793 Exempt. *Call for a Special Application in the Community of Temuco and Padre las Casas under Special Conditions for the Development of Projects of the Program "Thermal Conditioning for the New Plan of Atmospheric Decontamination in the Community of Temuco and Padre las Casas"*; Official Gazette of the Republic of Chile: Santiago, Chile, 2015.
10. National Energy Commission Chile. *Biomass Projects—Guide for Environmental Assessment of Non-Conventional Renewable Energies*; Technical Report; Ministery of Energy Chile: Santiago, Chile, 2007.
11. Vassilev, S.V.; Baxter, D.; Andersen, L.K.; Vassileva, C.G. An overview of the chemical composition of biomass. *Fuel* **2010**, *89*, 913–933. [CrossRef]
12. Kummamuru, B. *World Energy Resources: Bioenergy 2016*; World Energy Council: London, UK, 2016; p. 60. [CrossRef]
13. Institution for Diversification and Energy Saving Spain. *Biomass Energy*; Technical Report; Ministry of Industry, Tourism and Commerce: Madrid, Spain, 2007.
14. *Energy and Uses and Conservation Supply Curve Study in the Residential Sector*; Universidad de Concepción, Facultad de Ciencias Ambientales: Concepción, Chile, 2010; p. 404.
15. Hamins, A.P. Soot. In *Environmental Implications of Combustion Processes*; Puri, I.K., Ed.; CRC Press: Boca Raton, FL, USA, 1993; Chapter 3; pp. 71–95.
16. Simoneit, B.R.T.; Schauer, J.J.; Nolte, C.G.; Oros, D.R.; Elias, V.O.; Fraser, M.P.; Rogge, W.F.; Cass, G.R. Levoglucosan, a tracer for cellulose in biomass burning and atmospheric particles. *Atmos. Environ.* **1999**, *33*, 1–10. [CrossRef]
17. Flagan, R.C.; Seinfield, J. *Fundamentals of Air Pollution Engineering*; Dover Publications Inc.: Mineola, NY, USA, 2012; p. 557. [CrossRef]
18. Sippula, O. Fine Particle Formation and Emissions in Biomass Combustion. Ph.D. Thesis, University of Eastern Finland, Kuopio, Finland, 2010.
19. Ebeling, J.M.; Jenkins, B.M. Physical and chemical properties of biomass fuels. *Trans ASAE* **1985**, *28*, 898–902. [CrossRef]
20. Ministry of Health—Chile. *Code: Method CH-5G*; Technical Report 10; Public Health Institute—Chile: Santiago, Chile, 2012. [CrossRef]
21. Servicios Integrales de Calidad Ambiental Ltd. *Actualización del Inventario de Emisiones Atmosféricas paras las Comunas de Temuco y Padre Las Casas, Año Base 2017*; Technical Report; SICAM Ingeniería: Temuco, Chile, 2018.
22. AEN/CTN Aenor. *Residential Cookers Fired by Solid Fuel. Requirements and Test Methods*; Technical Report; Asociación Española de Normalización y Certificación: Madrid, Spain, 2002.
23. Neulicht, R.; Munoz, B.; Schaffner, K.; Parker, B. *Exploration of Emissions Factor Adjustments for Using Emissions Factors in Noninventory Applications*; Technical Report; RTI International: Durham, NC, USA, 2006.
24. Nussbaumer, T.; Czasch, C.; Klippel, N.; Johansson, L.; Tullin, C. *Particulate emissions from biomass combustion in IEA countries: Survey on Measurements and Emission Factors*; International Energy & Agency (IEA) Bioenergy Task 32: Paris, France; Swiss Federal Office of Energy (SFOE): Bern, Switzerland, 2008; p. 40.
25. Koyuncu, T.; Pinar, Y. The emissions from a space-heating biomass stove. *Biomass Bioenergy* **2007**, *31*, 73–79. [CrossRef]
26. Boman, C.; Nordin, A. Characterization of inorganic particulate matter from residential combustion of pelletized biomass fuels. *Energy Fuels* **2004**, *18*, 338–348. [CrossRef]

27. Werther, J.; Saenger, M.; Hartge, E.U.; Ogada, T.; Siagi, Z. Combustion of agricultural residues. *Prog. Energy Combust. Sci.* **2000**, *26*, 1–27. [CrossRef]
28. Roy, M.M.; Corscadden, K.W. An experimental study of combustion and emissions of biomass briquettes 456 in a domestic wood stove. *Appl. Energy* **2012**, *99*, 206–212. [CrossRef]
29. Habib, M.A.; Elshafei, M.; Dajani, M. Influence of combustion parameters on NOx production in an industrial boiler. *Comput. Fluids* **2008**, *37*, 12–23. [CrossRef]
30. Puri, I.K. *Environmental Implications of Combustion Processes*; CRC Press: Boca Raton, FL, USA, 1993; pp. 97–137.
31. Dupont, V.; Pourkashanian, M.; Williams, A.; Woolley, R. The reduction of NOx formation in natural gas burner flames. *Fuel* **1993**, *72*, 497–503. [CrossRef]
32. Chen, Y.; Shen, G.; Liu, W. Field measurement and estimate of gaseous and particle pollutant emissions from cooking and space heating processes in rural households, northern China. *Atmos. Environ.* **2016**, *125*, 265–271. [CrossRef]
33. Cooper, J. Environmental impact of residential wood combustion emissions and its implications. *J. Air Pollut. Control Assoc.* **1980**, *30*, 855–861. [CrossRef]
34. Hueglin, C.; Gaegauf, C. Characterization of wood combustion particles: Morphology, mobility, and photoelectric activity. *Environ. Sci. Technol.* **1997**, *31*, 3439–3447. [CrossRef]
35. Shen, G.; Xue, M.; Wei, S. Influence of fuel moisture, charge size, feeding rate and air ventilation conditions on the emissions of PM, OC, EC, parent PAHs, and their derivatives from residential wood combustion. *J. Environ. Sci.* **2013**, *25*, 1808–1816. [CrossRef]
36. Nussbaumer, T. Combustion and co-combustion of biomass: Fundamentals, technologies, and primary measures for emission reduction. *Energy Fuels* **2003**, *17*, 1510–1521. [CrossRef]
37. Nussbaumer, T.; Lauber, A. Formation mechanisms and physical properties of particles from wood combustion for design and operation of electrostatic precipitators. In Proceedings of the 18th European Biomass Conference and Exhibition, Lyon, France, 3–7 May 2010.

sustainability

MDPI

Review

The Potential Contribution of Decentralized Anaerobic Digestion towards Urban Biowaste Recovery Systems: A Scoping Review

Eftychia Ntostoglou *, Dilip Khatiwada and Viktoria Martin

Division of Energy Systems, Department of Energy Technology, KTH-Royal Institute of Technology, Brinellvägen 68, 10044 Stockholm, Sweden; dilip.khatiwada@energy.kth.se (D.K.); viktoria.martin@energy.kth.se (V.M.)
* Correspondence: ent@kth.se

Abstract: The potential contribution of decentralized approaches in implementing biowaste recovery systems has attracted interest in urban policy making and scientific research. Although the scientific literature on the topic is rapidly increasing, it is still limited and scattered. A comprehensive overview of current scientific knowledge is thus needed to support future research on decentralized options for biowaste recovery systems. Anaerobic digestion (AD) is a mature biowaste treatment technology that recovers energy and nutrients, and can close urban resource loops. Through a scoping literature review, this paper investigated decentralized AD and its potential contribution in implementing urban biowaste recovery systems. We identified opportunities and challenges for planning of decentralized AD, and concluded that these mainly concern: (a) digestate management; (b) the potential for local circularity with product valorization in outlets such as urban agriculture; and (c) the development and application of decision support tools. The findings highlighted the need to enhance scientific evidence on the impact of decentralized AD in different urban contexts. Results from published studies were highly context-specific, making it difficult to draw general conclusions. This study can support the transition to integrated planning of AD and wider urban biowaste recovery systems. Such planning must include a comprehensive analysis of configuration approaches.

Keywords: local circularity; decentralized biowaste management; circular economy; resource recovery; anaerobic digestion

Citation: Ntostoglou, E.; Khatiwada, D.; Martin, V. The Potential Contribution of Decentralized Anaerobic Digestion towards Urban Biowaste Recovery Systems: A Scoping Review. *Sustainability* **2021**, *13*, 13435. https://doi.org/10.3390/su132313435

Academic Editors: Luis M. López-Ochoa and José P. Paredes-Sánchez

Received: 5 November 2021
Accepted: 30 November 2021
Published: 4 December 2021

1. Introduction

Solid waste management is a pressing sustainability challenge for modern cities. Global and urban populations constantly grow, as does the amount of municipal solid waste (MSW) generated, with cities being accountable for approximately 70% of global waste [1]. In 2050, global waste is expected to reach 3.4 billion tonnes, a 70% increase compared to 2016 [2]. MSW is defined as the waste generated in municipalities, mainly composed of organic biodegradable waste, paper and cardboard, plastic, metal, and glass [2]. The growing MSW amount is largely attributed to the dominant linear model of global production and consumption that operates under a 'take-make-use-dispose' approach. This model is unsustainable, as resources are discarded after use and their value is lost [3,4]. The circular economy (CE) has gained increasing attention as a means to rethink overall resource management: it aims to preserve the value of products, materials and resources for as long as possible and minimize waste generation [5]. There are significant opportunities to transition from linear to circular resource management and apply CE as a transition strategy to sustainable low- and zero-carbon societies [6–8].

According to the 'waste hierarchy' framework, waste prevention should be the top priority of strategies towards sustainable resource management. Nonetheless, resource recovery is also an indispensable component of the hierarchy as a strategy to manage unavoidable waste [9–11]. The terms 'waste recovery' or 'resource recovery' describe any process that uses waste as input to replace resources (e.g., extraction of virgin resources) that would be used otherwise [12,13]. Resource recovery captures value in the system that would otherwise be lost. In this context, the establishment of effective resource recovery systems is essential, and has attracted increasing scientific and public interest [3].

1.1. Urban Biowaste Recovery: A Largely Untapped Potential

Biowaste has a crucial role in the implementation of resource recovery and wider sustainability transitions through CE [8,14]. 'Biowaste' was here considered as the organic fraction of municipal solid waste (OFMSW): food and kitchen waste from households and institutional and commercial (including restaurants and food markets) buildings, and comparable waste from food processing plants, as well as green waste from parks, yards, and green spaces [15]. It usually constitutes the largest fraction of municipal solid waste [2]. In this paper, 'biowaste recovery' refers to resource recovery from biowaste streams. Biowaste recovery systems can contribute to the development of a circular bioeconomy: an economy in which biowaste and other bioresources are used in bioenergy and biorefinery systems to generate high-value biobased products, such as biofuels for energy services and nutrient-rich biofertilizers [13,16].

Nevertheless, biowaste still remains a largely untapped resource globally. In most urban areas, it is still collected while mixed with other MSW types and disposed in landfills and open dumps [2,17]. The current global status of biowaste management highlights the shortcomings of the linear economic model and the largely unexploited potential to close resource loops through biowaste recovery [18]. The authors of [19] conducted a scoping review on circular organic waste management. Among key future research directions, they suggested to explore different pathways for biowaste recovery, with focus on assessing the value added through energy and nutrient recovery [19].

1.2. Anaerobic Digestion (AD): A Pathway for Biowaste Recovery

Four main types of technologies can treat biowaste: (a) direct use (e.g., direct combustion); (b) biochemical treatment (e.g., fermentation and anaerobic digestion); (c) physicochemical treatment (e.g., transesterification); and (d) thermochemical treatment (e.g., gasification) [20]. Figure 1 shows that some treatment technologies recover bioenergy among their products—the so-called "waste-to-energy" (WtE) technologies. WtE options address two global challenges simultaneously: the growing amount of waste, as well as the increasing energy demand, providing a clean energy alternative to replace fossil fuel use. They contribute to 'cleaning' and diversifying the energy mix and reducing reliance on external energy imports towards resilient energy systems [21]. In addition, some WtE technologies provide opportunities to recover nutrients and close loops in bioresource management [22].

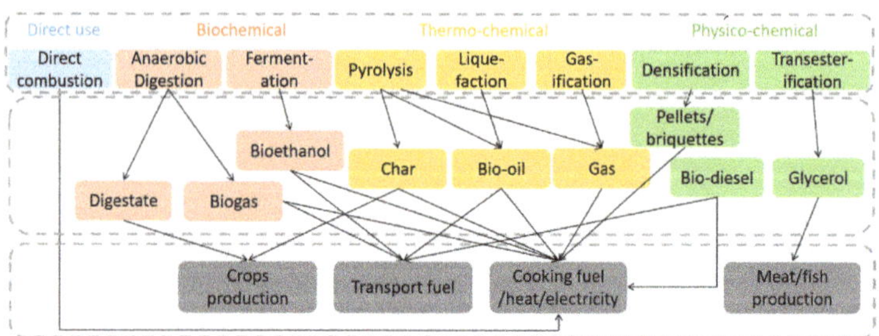

Figure 1. WtE pathways for biowaste treatment. Adapted from [20].

Among available WtE technologies, anaerobic digestion (AD) is an established technology: there are various AD options with a high technology-readiness level that are commercially available and applied [23]. AD is a biochemical process that decomposes organic matter in the absence of oxygen. It can treat biowaste and other biomass feedstocks such as agricultural residues, livestock manure, wastewater and fecal sludge, industrial waste, and energy crops [8,24]. The products of the process are biogas (a renewable energy source) and digestate, which contains organic matter and nutrients such as nitrogen, phosphorus, and potassium (N,P,K) [25]. Biogas can be used as cooking fuel or converted to heat and electricity through combined heat and power (CHP) engines. It can also be upgraded to biomethane to use as vehicle fuel or to inject into natural gas grids [26]. Through biogas combustion, AD systems avoid methane emissions compared to other conventional biowaste management options such as landfilling. In turn, digestate can be further processed to use as fertilizer, soil amendment, or livestock bedding [20]. Another emerging option is digestate use in 'digeponics', a type of hydroponics in which digestate-based products are used as substrate to grow plants [27]. Overall, AD has a multifunctional character [28], and can contribute to sustainable and circular resource management towards energy and food security, waste management, and sanitation [29,30]. The authors of [31] considered AD '*not as an energy technology but as a technology that addresses challenges across multiple resource domains*'. This multifunctional character requires an integrated approach for AD planning: in this study, the term 'integrated' refers to the assessment of direct or indirect interlinkages across scales, systems, and sectors.

Figure 2 depicts the AD process chain in the urban biowaste recovery context. As shown, the process chain comprises three key stages: the substrate chain, AD treatment, and the product chain. At the substrate chain stage, various urban sources generate biowaste. Urban biowaste is collected and transported for treatment. Pretreatment can enhance substrate quality. Then, the substrate undergoes treatment (i.e., hydrolysis, acidogenesis, acetogenesis, and methanogenesis) through a suitable AD technology. Biogas and digestate products can undergo post-treatment, depending on the intended product use (e.g., biogas upgrade to biomethane and use as transport fuel). Finally, the end products are stored and distributed for valorization at suitable outlets [32].

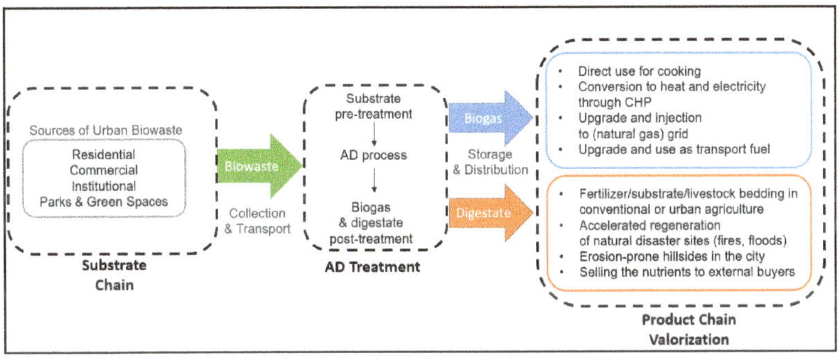

Figure 2. Process chain of anaerobic digestion for urban biowaste recovery. Based on [32,33].

Despite the potential benefits, AD implementation for urban biowaste recovery is still low compared to its full potential at the global level [34,35]. According the International Energy Agency (IEA), actual biogas production amounted to 35 million tonnes of oil equivalent (Mtoe) in 2018. This represents roughly 6% of the full biogas potential (i.e., 570 Mtoe) using available feedstocks. Realizing the full potential could supply approximately 20% of the world's current gas demand [36]. The feedstocks leading to this estimate included crop residues, livestock manure, MSW, and wastewater. MSW represents roughly 20% of the global biogas potential (i.e., 112 Mtoe). Therefore, increasing AD implementation can significantly contribute to achieving the biowaste recovery potential. Further research on the implementation of urban AD systems, as well as related opportunities and challenges, can support decision making. In this paper, the terms 'urban AD systems' and 'urban AD' refer to AD systems that treat urban biowaste as primary feedstock.

1.3. Centralized and Decentralized Approaches for Integrated Biowaste Management Systems

Modern cities have mainly followed centralized approaches to organize waste management, as well as other resource management systems such as for energy and water [37]. Drivers of centralized approaches include economies of scale and transport costs [38]. However, the rapidly growing global waste generation pushes for renewal and expansion of relevant infrastructure; e.g., disposal and treatment facilities, waste collection fleets, etc. This infrastructure demand puts increasing pressure on centralized waste management systems and can trigger sustainability challenges [39]. For example, collection fleets must travel larger distances to treatment and disposal facilities, using larger amounts of vehicle fuels (economic cost) and thus increasing transport-related greenhouse gas (GHG) emissions (environmental cost) [18]. MSW collection and transport require up to 40% of municipal revenues for cities in developing countries [40]. In this context, decentralized systems are an alternative to shape waste management systems, and have attracted interest by practitioners, policy makers, and the scientific community [41].

Considering relevant scientific literature, the authors of [42] applied bivariate analysis on different types of energy technologies and found various potential benefits of decentralized, small-scale technologies (e.g., faster diffusion, opportunities to escape lock-in) to facilitate decarbonization. In the bioenergy context, the authors of [43] reviewed international case studies and identified opportunities and challenges for the implementation of decentralized bioenergy systems. They mapped interlinkages across the three sustainability pillars (economic, social, and environmental), and highlighted market viability as a major challenge. To support product establishment in the market, the authors emphasized opportunities to integrate bioenergy production with other sectors to develop closed-loop systems. However, the authors of [43] did not consider urban biowaste among the feedstock types addressed (forestry and agricultural residues, livestock manure).

The implementation of urban AD has followed a pattern similar to many waste management systems, as most operating urban AD systems globally are centralized [44]. It has been suggested that partial decentralization of biowaste management can better support the transition from linear to circular systems, and shift perspective from 'waste' management to a wider resource management approach [45]. Nevertheless, the authors of [37] characterized the current scientific literature on technologies for decentralized biowaste treatment *'fragmented and incomplete'*. To begin filling this gap, they classified and compared decentralized options for urban biowaste treatment (including AD) through extended material flow analysis (EMFA). Their analysis focused on techno-economic aspects of decentralized options and did not investigate system level planning. Through an interview-based stakeholder analysis, the authors of [41] identified institutional drivers and barriers towards the implementation of decentralized biowaste management systems. However, to the authors' knowledge, there was currently no systematic and comprehensive review to provide an overview of decentralized AD at the system level addressing questions such as: What is the current scientific knowledge on key planning aspects for decentralized AD at system level? What are the opportunities and challenges for planning of decentralized AD? What is the potential contribution of decentralized AD towards the implementation of urban biowaste recovery systems? Moreover, relevant studies have rarely provided explicit definitions of centralized and decentralized management systems.

1.4. Paper Objective and Outline

Through a scoping literature review, this study aimed to provide a comprehensive overview of current scientific knowledge on decentralized AD. It focused on its potential contribution towards the implementation of urban biowaste recovery systems. Such systematic assessment of scientific knowledge related to decentralized AD is currently not available in the literature. The synthesis led to key opportunities and challenges of decentralized AD planning. These can guide future decision making and scientific research on planning of decentralized AD and urban biowaste recovery systems.

The rest of the paper is organized as follows: Section 2 outlines the research design. Section 3 synthesizes current knowledge on decentralized AD for urban biowaste recovery based on the scientific literature. Section 3.1 presents definitions of centralized and decentralized approaches. Through a literature classification, Section 3.2 discusses key planning aspects of decentralized AD systems for urban biowaste recovery. Section 4 further investigates emerging research themes that can guide the implementation of decentralized AD. Section 5 summarizes key messages through the lens of opportunities and challenges for future development of decentralized urban AD. Finally, Section 6 draws the study conclusions.

2. Research Design

A study must follow a transparent and systematic process to employ a literature review as a robust research method [46]. This study conducted a scoping review with the aim to analyze emerging evidence and research gaps on the topic, as well as investigate how research is conducted [47]. It was based on the Preferred Reporting Items for Systematic Reviews and Meta-Analyses (PRISMA) standard [48], an established approach to guide scoping and systematic reviews.

The literature review process followed these steps: (1) using a search string to identify relevant scientific studies through two scientific databases; (2) screening these studies based on a series of eligibility criteria; and (3) conducting a qualitative synthesis. The qualitative synthesis exposed key opportunities and challenges of decentralized AD through: (a) provision of definitions of centralized and decentralized approaches; (b) classification of current knowledge on decentralized AD; and (c) identification of emerging themes that can support future research. Figure 3 visualises the review process (data related to the review material are also provided in Tables S1–S5 of the Supplementary Material).

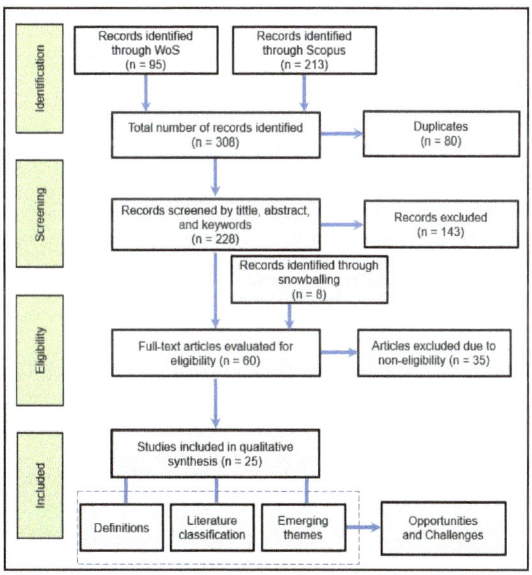

Figure 3. The literature review process, based on [48].

2.1. The Literature Collection Process

The following search string was used (using Boolean operators and truncation) to identify relevant literature in the scientific databases 'Scopus' and 'Web of Science' (WoS):

("anaerobic digestion" OR "biogas" OR "digestate" OR "nutrient") AND ("decentrali*" OR "small" OR "micro" OR "centrali*" OR "large") AND ("urban" OR "city" OR "cities") AND ("biowaste" OR "organic waste" OR "organic solid waste" OR "organic fraction of municipal solid waste" OR "food waste" OR "food and garden waste")*

The search string was applied in a search based on title, abstract, and keywords. The search was limited to peer-reviewed journal publications in English, in the time range of 2010 to July 2021. It was assumed that current literature reflected and incorporated all major scientific knowledge from studies published before 2010. Duplicate results from the two databases were removed.

2.2. Eligibility Criteria to Screen the Review Material

To screen the remaining (unique) documents, a series of eligibility criteria were set (see Table 1). In the first screening round, the eligibility criteria were applied by reading the title, abstract, and keywords of each document. In the second screening round, each document was read in full to determine eligibility based on the same criteria. Several papers were excluded based on the second criterion: while they addressed AD for urban biowaste, they often focused on AD at the plant scale and technical/operational aspects of the AD process. In addition, 'snowball sampling' [49] was also applied: some studies identified through database search cited or were cited by papers relevant to the review criteria (screening stage, see Figure 3). A final list of 25 records (20 original research articles and 5 review articles) was compiled to conduct qualitative synthesis. Twenty additional references from the wider scientific literature were cited to provide background to the analysis, where appropriate (e.g., [50]).

Table 1. Eligibility criteria and relation to classification parameters.

	Eligibility Criteria for Paper Screening (Methodology)	Relevant Classification Categories (Analysis)
1	Does the paper consider urban biowaste (and subcategories) as the primary feedstock for anaerobic digestion?	• Feedstocks • Geographic scope of analysis • Embeddedness in the urban environment
2	Does the paper address decentralized approaches, or compare or combine AD configuration approaches?	• Configuration approach
3	Does the paper address aspects of AD planning at the system level (as a pathway for urban biowaste recovery) beyond the plant/project level?	• Selection of treatment technologies • Implementation aspects • Methodological tools • Circular (bio)economy

The selected eligibility criteria were based on (necessary) assumptions and could have been subject to bias, despite all efforts for objectivity. Nevertheless, the research design was reported fully, and it can be replicated or modified by other researchers for future research (see Supplementary Material).

2.3. Qualitative Synthesis

The qualitative synthesis was based on literature classification and thematic analysis [51,52]. The literature classification approach was informed by [7,19]. Eight classification categories were formed based on thematic analysis of the literature and the eligibility criteria. Table 1 shows how the classification categories (analyzed in Section 3.2) related to the eligibility criteria. The literature classification provided an initial organization of the review material and set the ground for further thematic analysis. Thematic analysis is a research method that aims to identify, analyze, and report patterns (themes) within data [53]. It was used to synthesize the review material.

3. Current Scientific Knowledge on Decentralized AD for Urban Biowaste Recovery

The literature review findings are presented in two sections aiming to discuss: definitions of centralized and decentralized systems (Section 3.1) and highlight key planning aspects of decentralized urban AD (Section 3.2).

3.1. Definitions of Centralized and Decentralized AD Systems

To explore centralized and decentralized AD approaches, the authors first examined how they are defined in the papers reviewed. Table S6 (see Supplementary Material) summarizes definitions/descriptions of centralized and decentralized approaches used in the 25 papers reviewed. The findings showed that definitions of 'centralized' and 'decentralized' treatment and 'large-scale' and 'small-scale' facilities could largely vary depending on the system boundaries and contextual characteristics of each study.

In most cases, the authors did not provide explicit definitions of centralized and decentralized AD systems. Several studies described decentralized systems as: (a) consisting of small-scale plants and (b) located close to the waste source, compared to centralized approaches. For example, according to [54], decentralized systems consist of small-scale AD plants that are approximately *'75 m^2 (15 m × 5 m) to accommodate all the equipment and the required space around'*. In contrast, the authors of [37] describe decentralized treatment systems as: *'A class of treatments that encompasses relatively small facilities capable to metabolize less than approx. 10 tonnes biowaste/year'*.

In turn, there is discrepancy in definitions of 'large-scale' and 'small-scale'. The papers reviewed used different parameters to define small-scale AD (see Table S6 in Supplementary material). Some studies used digester volume to distinguish small- and large-scale systems (e.g., [55]), while others referred to treatment capacity (e.g., [56]), biogas production (e.g., [33]), or installed capacity (e.g., [57]). Even when two studies used the same parameter to describe scale, the ranges set could largely vary. For example, in a local context with large feedstock (e.g., biowaste) availability, the threshold for 'small-scale' could be set higher compared to a geographic area with lower feedstock availability. The large variety of definitions highlighted that planning of AD systems requires integrated approaches based on contextual characteristics of the geographic area of implementation [58].

For this study, 'decentralized systems' consisted of relatively small-scale facilities usually located at short distance from waste sources and end users. In contrast, 'centralized systems' consisted of (usually fewer) large-scale facilities located at larger distance from the city. Based on the review material, one example of centralized AD was an urban biowaste large-scale facility 130 km from the city of Brussels that treats food waste with capacity of 50,000 tonnes/year [17]. One example of decentralized AD was a system of 170 small-scale AD sites to treat urban biowaste in the Lyon metropolitan area, each located at a maximum distance of 5 km from waste sources and with capacity of less than 61 tonnes/year [54].

3.2. Key Planning Aspects of Decentralized AD for Urban Biowaste Recovery

To analyze the scientific knowledge on planning of decentralized AD for urban biowaste recovery, the review material was classified under eight categories. In the Supplementary Material, Table S7 lists the eight categories and their subcategories. Table S8 shows the paper classification. Key findings from this analysis are discussed below, and were organized according to the eligibility criteria for paper selection (see Table 1).

3.2.1. Urban Biowaste as Primary Feedstock for AD

The authors screened for papers that focused on urban biowaste as exclusive AD feedstock or together with other feedstocks. This criterion included papers that focused on urban biowaste subcategories, such as food waste. The papers were classified while considering the feedstock types they addressed, their geographic scope of analysis, and whether/how they addressed AD embeddedness in the urban environment.

Feedstocks

AD can treat a wide range of biomass feedstocks. Even with focus on one bioresource category (here urban biowaste), AD feedstock characteristics may largely vary from one study to another, depending on various aspects. Firstly, biowaste must be separated from other MSW fractions as early as possible in the AD process chain. Early sorting and collection maximize feedstock quality, which in turn largely determines treatment efficiency and the quality of the final AD products [41]. The transition to separate biowaste sorting and collection has been addressed by other studies (e.g., [59,60]) and is beyond the scope of this paper. At times, biowaste subcategories (food and green waste) were collected and treated combined or separately. For example, four papers focused on food waste as key AD feedstock, while [17] found that combined collection of food and green waste, separated from other waste fractions, was the most efficient collection approach for their Brussels case study. The suitable feedstock, as well as the suitable mode of separation and collection, were context-specific for each recovery project, depending on several characteristics of the city studied. For example, the authors of [61] studied the biogas performance of urban feedstocks collected through different methods (e.g., mixed collection and mechanical separation, separation at source, and hand sorting), and highlighted that urban characteristics such as morphology (e.g., urban density and size of the streets) largely influenced waste generation, the sorting and collection methods, and thus treatment efficiency. Moreover, biowaste can undergo pretreatment (see Figure 2) to enhance treatment efficiency; e.g., processing through a chopper mill and feeding to the digester through pumps [62]. Biowaste can be codigested with other substrates such as sewage sludge: five reviewed papers addressed codigestion. In the review material, 13 papers focused on urban biowaste as main AD feedstock, which is also termed 'organic fraction of municipal solid waste (OFMSW)' or 'food and green waste'. Three other papers considered the wider waste management system (including nonorganic MSW) and addressed biowaste treatment among other waste categories. The classification of feedstocks showed that the scientific literature has addressed several feedstock options in the decentralized context. Feedstock characteristics are highly influenced by several context-specific factors that must be assessed in AD planning (e.g., optimization and sustainability assessment).

Geographic Scope of Analysis

System boundaries can largely vary in terms of how studies approach the 'urban' scale of analysis: a paper may study urban biowaste recovery at the municipal, metropolitan or even regional level. For example, the authors of [63] compared aspects of environmental and economic performance of different treatment systems between municipal districts in the metropolitan region of Porto, Portugal. The authors of [64] developed a DST for AD planning at regional/county level, while those of [38] applied a multilevel analysis considering the deployment of biowaste recovery systems at the national, district, and organizational level. The study in [56] focused on a small community (land area: 80 km^2, 17,000 inhabitants, 550 kg waste per capita per year).

Moreover, it was observed that most papers reviewed mainly analyzed case studies in developed regions. This tendency reflected the geographic distribution of implemented decentralized AD projects, which were also found mainly in developed countries. While decentralized AD projects in developing contexts have also been addressed by the scientific literature and implemented in practice, these were mainly found in rural areas, and often focused on other feedstock types such as livestock manure (e.g., [65,66]).

Embeddedness in the Urban Environment

Accounting for potential interactions between AD systems and their geographical space of implementation is essential to achieve integrated planning. The study in [67] indicated that the embeddedness process differs between the rural and urban context, depending on the socioeconomic structures in each context. Moreover, decentralized urban

AD presents opportunities to locate treatment plants within the urban environment, in contrast to centralized facilities, usually located in city outskirts. In such cases, AD systems must be embedded within city boundaries, in harmony with the urban environment [68]. Essential questions to address include: Where will AD facilities be located? How do they affect pre-existing urban elements (e.g., other types of infrastructure)? The authors of [18] indicated that, in the shift from centralized to more decentralized, AD systems can be embedded at various urban levels: an AD plant may target biowaste at the building, district, or municipal level. It is thus essential to address different embeddedness levels. Including [18], 18 papers addressed different levels of AD embeddedness in the urban environment, either implicitly or explicitly. For example, the authors of [55] focused on embeddedness at the building level, and [45] at the municipal level. Some papers referred to more than one level of embeddedness. For example, the authors of [62] addressed embeddedness at the building level (community café) as well as district level. In their London case study, households close to the plant provided the AD feedstock. The biogas produced was used for the plant's energy needs (heat and electricity), but also by a nearby community café (cooking fuel). Finally, nine papers considered the potential for urban AD embeddedness through synergies with UA. The reasoning for such an integration was to use recovery products in UA, thus developing local circularity. This potential integration is further analyzed in Section 4.

3.2.2. Configuration Approaches for AD Planning

AD systems can be based on centralized, decentralized, or combined approaches that mix centralized and decentralized configurations. The system's configuration can largely influence the quality of AD products [41] and the wider sustainability impact of the system. In the review material, 11 papers focused only on decentralized approaches for AD planning. One study focused on centralized approaches. One study addressed AD planning without referring to a specific configuration approach. Finally, 12 studies compared different configuration scenarios and combined approaches to enhance the performance of biowaste recovery systems. These studies emphasized that the effect of different system configurations of AD and wider urban biowaste recovery systems has been marginally addressed. Moreover, several studies highlighted potential opportunities of implementing decentralized urban AD. However, the description of such opportunities was rarely supported by relevant scientific evidence. The findings showed that quantitative and qualitative assessments of opportunities, as well as related challenges (as part of sustainability assessment), are limited. Future research is essential to further analyze and integrate system configurations into planning of AD and wider urban biowaste recovery systems. Sections 4 and 5 further address pathways to enhance knowledge of decentralized urban AD.

3.2.3. Other System Level Aspects for AD Planning

This paper addresses decentralized AD at the system level as opposed to the plant level. Plant-level studies refer to those that focus on individual AD facilities, mainly addressing technical/operational and technoeconomic aspects of the AD process itself. In contrast, the screening aimed for system-level studies that focused on networks of AD facilities to treat a city's biowaste or a fraction of it. Papers were classified considering treatment technologies that can be combined with AD to develop biowaste recovery systems. Other classification categories were 'methodological tools' and 'implementation aspects', because system-level analyses need to address various implementation aspects and can use a wide range of methods. Moreover, the sustainability context under which AD was addressed was also classified, with focus on the contexts of circular economy and bioeconomy.

Selection of Treatment Technologies

The treatment technologies addressed in each paper were also classified. While this review focused on AD, several technologies for biowaste treatment exist (see Figure 1).

Sixteen papers reviewed focused only on AD as an option for biowaste recovery. Nine papers compared and combined AD with other treatment technologies to select suitable technology combinations in different contexts. For example, the authors of [63] assessed the economic and environmental costs of different technology scenarios (landfilling with gas recovery, centralized incineration, centralized AD, and centralized and decentralized composting) using life-cycle assessment (LCA) and spatial analysis. Their results showed trade-offs occurring in each scenario (e.g., local composting had low economic costs, but high environmental costs compared to AD), and that system design must be guided by urban characteristics (e.g., local composting is suitable for remote, less dense neighbourhoods). Moreover, the authors of [38] assessed scenarios of partially and completely decentralized configurations of AD and gasification. Both AD and gasification scenarios led to sustainability benefits. The gasification scenario had the highest economic (e.g., revenues from upgrading biogas to fuel), environmental (e.g., GHG emission savings), and social (e.g., job creation) benefits. However, gasification options are still less advanced than AD at the technical and commercial level [38]. The findings above showed the complexity of identifying suitable technologies and configurations as part of planning biowaste recovery systems. Moreover, planning such aspects must reflect the case study's specific local context. Therefore, additional case study research investigating different scenarios can better inform scientific knowledge and decision making regarding AD and biowaste recovery systems. Furthermore, the findings of [38] highlighted AD as a low-hanging fruit: a conventional technology ready to implement while other novel technologies for high-value bioproducts are further developed.

Implementation Aspects

The need for integrated planning across the three sustainability pillars has been highlighted in the wider literature for resource recovery systems [69]. The reviewed studies addressed various AD implementation aspects. The level of analytical detail, system boundaries, methodological tools (see next section), and metrics used largely varied across studies. Some papers did not provide in-depth analyses, but instead merely provided preliminary descriptions of implementation aspects. For example, while 12 papers addressed environmental aspects of AD implementation in their analysis, only 4 examined environmental performance comprehensively through LCA approaches. Other studies followed less-comprehensive approaches to assess environmental performance. For example, the authors of [64] estimated expected CO_2 emissions based on distance of biowaste collection and transportation. The study of environmental performance is further addressed in Section 4.2.

Twenty papers included technical aspects (e.g., digester sizing) to assess implementation. Sixteen papers considered economic (e.g., capital and operational costs), twelve considered environmental (e.g., associated GHG emissions), and nine addressed social (e.g., plant acceptance) aspects. Moreover, seven papers addressed institutional aspects for AD implementation. For example, the authors of [38] assessed (national) policy frameworks to identify challenges of the current waste management system, for the case of the United Kingdom. The authors of [70] mentioned that policy interventions such as grant incentives can impact AD uptake significantly. Spatiotemporal aspects are also crucial to consider for AD planning [71,72]. Seven papers addressed spatial dimensions, while only two papers addressed both spatial and temporal aspects. The study in [64] used spatial analysis to develop an agent-based model (ABM), and included the temporal rate of implementation (slow, mid, or aggressive) as a parameter to design AD planning scenarios. The authors of [18] used spatial analysis and considered a range of time periods (over 1, 10, and 15 years) to estimate AD implementation over space and time. Overall, comprehensive assessment of all relevant implementation aspects can enhance the level of detail and accuracy of AD planning.

Methodological Tools

Integrated AD planning requires support from adequate methodological tools. The classification shows that the reviewed studies applied a wide range of methodological approaches, from material flow analysis (MFA) (four papers) to multilevel perspective (MLP) [45] and visual analytics [64]. Most studies applied mixed research methods, and as mentioned for 'Implementation aspects', analytical depth largely varied. For example, most papers employed some form of literature review as part of their studies: an overview of current knowledge and research gaps for research motivation. Five papers employed literature review as their main methodological tool, but only [57] followed a systematic process. The other papers applied more ad hoc approaches in using a literature review as a key research method, although some supported their review findings with interviews with experts. None of these papers provided a systematic and comprehensive analysis of urban AD planning through decentralized approaches. Notable methodological approaches that can support further research towards integrated planning of AD and wider biowaste recovery systems are discussed further in Section 4.

Circular Economy and Bioeconomy

The role of AD as part of a circular economy and circular bioeconomy is also crucial to consider as an aspect of integrated planning. In the review material, only three papers addressed urban AD explicitly in the context of circular bioeconomy. Namely, the authors of [17] mentioned the study of bio-based materials deriving from AD as a future research step. The authors of [57] reviewed digestate valorization options, including advanced technologies, to support the development of biorefinery systems. The authors of [41] identified institutional factors that enabled or constrained implementation of decentralized biowaste management systems in the context of the circular bioeconomy. Therefore, potential interlinkages between urban AD and other stages in cascading biomass use were beyond the scope of these studies and have not been researched explicitly. However, there are several options to cascade biomass use (e.g., production of chemicals through advanced biorefineries), beyond energy and nutrient recovery through AD biogas and digestate. It is thus crucial to integrate AD in the wider context of circular bioeconomy comparing it to other options. Moreover, 16 papers contextualised their contribution within circular economy, while 6 papers referred to a wider sustainability context.

4. Emerging Research Themes on Decentralized AD towards Integrated Planning of Urban Biowaste Recovery

Decentralized AD can enable local circularity or '*local valorization loops*' by minimizing distances along stages of the value chain [54]. Contrary to centralized approaches, in decentralized systems, waste is treated closer to the source and end users, and thus recovery products can find local end uses, adding value to the urban environment. Figure 4 provides an example in which decentralized AD contributes to a local valorization loop: biowaste is collected from sources such as households and UA and treated through decentralized AD. Biogas produced can be used by the households; e.g., as cooking fuel, and digestate as fertilizer in UA.

The potential for local circularity must be assessed to plan such localized systems (feedstock sources, treatment methods, product outlets, and end uses). In this context, assessing the relation between resource supply and demand is crucial. Here, 'supply' refers to AD products and 'demand' to the resource requirements of potential product outlets locally. The authors of [63,73] used the term '*urban sinks*' to describe local product outlets and '*local sink capacity*' when referring to the potential to use recovery products locally. They highlighted that scientific knowledge is limited concerning local sink capacity and the performance of decentralized treatment systems under large-scale deployment.

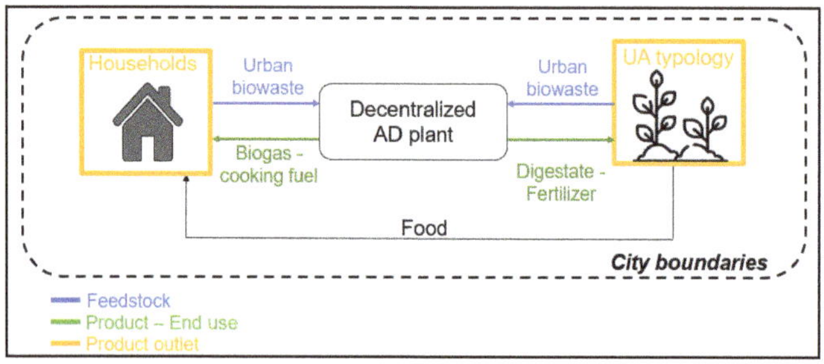

Figure 4. Example of a simplified 'local valorization loop' (own graph). Icons used were designed by Freepik (https://www.freepik.com, accessed on 2 July 2021).

Building on Section 3, this section identifies emerging research themes on decentralized AD with focus on integrated planning of urban biowaste recovery systems. The findings showed that all reviewed papers addressed biowaste recovery through local valorization loops either explicitly or implicitly. The emerging research themes identified were: (a) spatial analysis; (b) life cycle analyses of environmental performance; (c) decision support tools (DSTs) and frameworks; (d) nexus approaches; and (e) UA as an entry point for embeddedness in the urban environment. Most papers touched upon more than one of these themes (see Table S9 in the Supplementary Material). For example, the authors of [64] developed a DST for AD planning (c), largely based on spatial analysis (a), while it also adopted a nexus approach (d) for sustainability assessment. Each theme is further discussed below.

4.1. Spatial Analysis

The integration of spatial dimensions in planning of resource systems has received significant scientific interest [50]. This interest is reflected in the decentralized AD context. Spatial analysis combines methods of mathematical optimization and geographic information systems (GIS) to determine the spatial organization of AD systems. It informs decisions to achieve balance between supply and demand considering quantity, quality, and availability of biowaste feedstock, treatment facilities, product outlets, and their spatial distribution. For example, the authors of [54] presented a spatial optimization model to support the design of decentralized urban AD systems. For the case of Lyon, they found sufficient digestate potential to complement nutrient demand in periurban agriculture (PUA). For biogas, they assumed conversion through CHP, but did not provide further analysis; e.g., comparison to local electricity and heating demands. Moreover, their model focused on minimization of payload distances, while assessment of other aspects (e.g., environmental, social) was proposed for future work. Using spatial optimization, the authors of [73] explored the energy impact of upscaling UA and the potential to cover UA's energy and nutrient demand through biowaste recovery via AD, composting, and insect rearing. For the cases of Lyon and Glasgow, they found that digestate supply far exceeded UA's nutrient demand. Additional urban outlets would be required to use digestate surplus locally. Biogas use through CHP could cover only part of UA's energy demand (heat and electricity). On the other hand, they found that waste valorization contributed to reducing UA's carbon footprint (−7.9% for Glasgow, −12.6 for Lyon compared to upscaled UA scenarios without waste valorization in UA). The study in [63] applied spatial analysis to compare the environmental performance of municipal districts, as well as their sink capacity and related logistics (see Section 4.2); the authors of [18,64] also applied spatial analysis to develop DSTs (see Section 4.3).

Findings of spatial analysis supported that decentralized AD could contribute to local circularity with positive sustainability impact. Spatial methods assessed local sink capacity and informed decisions to match digestate supply with local fertilizer demand and plan for surplus management while considering sustainability implications [73]. Biogas management presented fewer technical challenges than digestate management. Current spatial methods also highlight the complexity of AD planning that integrates decentralized approaches. It is essential to further study UA and other urban outlets for product valorization. Local data are essential, but may be difficult to obtain due to lack of access and/or documentation. Future research can further develop/refine spatial methods to address different configuration approaches.

4.2. Life Cycle Analyses of Environmental Performance

The environmental performance of decentralized AD systems remains largely unclear. As mentioned in Section 3.2, few studies have examined environmental performance comprehensively through LCA. The authors in [56] conducted LCA for biowaste treatment scenarios for a small Italian community. They considered three impact categories: global warming potential (GWP); acidification, eutrophication, and ozone depletion; and photochemical ozone creation. Decentralized AD with digestate composting was the best-performing scenario for all impact categories. However, they also highlighted that AD environmental performance was largely dependent on local digestate use. The system's emissions may increase significantly if digestate requires transport for treatment and use elsewhere, largely due to transport-related GHG emissions. The authors of [17] conducted an LCA of centralized and more localized biowaste recovery systems and found that, under certain conditions, treatment systems located closer to the city performed better while considering the endpoint categories: human health, ecosystem damages, and availability of resources. The local AD scenario had the best environmental performance in their Brussels case study. The scenario referred to a large-scale plant (capacity: 50,000 tonnes/year) due to lack of local data for small-scale AD. It was located within the metropolitan area, compared to other scenarios, with a centralized AD facility located 130 km outside Brussels. Only their composting scenario included local small-scale facilities, which performed well in terms of resource use, but had lower overall environmental performance compared to local AD. The authors addressed the lack of other decentralized scenarios as a study limitation; they highlighted the need to expand research to assess the variety of decentralized biowaste treatment scenarios. The authors of [63] assessed economic and environmental metrics (annualized treatment cost and GWP) of centralized and decentralized treatment scenarios, while also considering spatial parameters such as urban sink capacity and related logistics (allocation of compost bins and urban farms as product outlets) across municipal districts of the Porto metropolitan area, applying LCA and spatial analysis. Local composting had the lowest economic costs and centralized AD the lowest environmental costs. They also found that additional urban farms would increase local sink capacity, but marginally reduce economic treatment costs (range of 0.5–2.5%). Environmental savings largely varied across the municipalities considered (range of 0.1–39.9%). The authors of [63] highlighted that their results largely varied due to the influence of context-specific factors such as urban density (see also Section 3.2.1), the energy sources of the electricity grid, and the potential for local digestate use. Finally, the authors of [74] conducted LCA while addressing not only AD configurations, but also various end uses for each configuration assessed. For the case of Singapore, they found that all AD scenarios performed better than incineration. Considering 17 impact categories, centralized AD for transport fuel and decentralized AD for cooking fuel were the scenarios with the highest environmental savings, with the latter performing best in terms of GWP and fossil fuel depletion.

Similar to studies of spatial analysis, studies with a focus on environmental performance highlighted the importance and challenges of assessing local sink capacity and digestate management. Further research can enhance knowledge on environmental implications of decentralized applying comprehensive methodological approaches, such as LCA and studying a variety of decentralized scenarios. Current findings highlighted AD's context-specific nature, and hence the need to conduct environmental-impact assessments that integrate local characteristics.

4.3. Decision-Support Tools and Frameworks (DSTs)

As described above, spatial analysis and LCA of environmental performance are valuable methodological tools to support AD planning. However, planning decisions such as the selection of suitable configurations also require more overarching decision support tools and frameworks (DSTs) [74]. DSTs are often based on multicriteria decision analysis (MCDA), which aims to address all relevant AD implementation aspects (see Section 3.2) while considering the three sustainability pillars and context-specific characteristics. DSTs that apply MCDA can thus support decision making towards integrated AD planning.

Methods such as spatial analysis and LCA can be part of such overarching DSTs. For example, the authors of [18] used spatial optimization as well as LCA data as part of their DST to estimate GWP. Their DST used modeling of MSW distribution, optimization of the management system (with focus on cost optimization), and a multicriteria framework for sustainability assessment. Their findings showed that, compared to the conventional MSW scenario (incineration), the combined centralized/decentralized AD system could double electricity profits through biogas, reduce capacity land fragmentation by 75% (thus enhancing land use) and GWP by 19%, operational expenses up to 50%, and the required transport fleet up to 15%. The authors recommended future research on the effect of different planning priorities (e.g., prioritizing GWP performance over economic costs) to assess centralized and decentralized approaches. The authors of [38] developed the Systems Thinking Approach to Resource Recovery (STARR) framework based on the case of the United Kingdom. It included a review of national waste management policy and a multilevel system analysis, which applied MFA to measure the potential recovery at the national, community, and organizational (supermarket) level. At the community level, the authors developed three scenarios for biogas and digestate production, and conducted sustainability assessment that included economic, environmental, and social parameters and indicators (see also Section 3.2). The authors of [64] developed an ABM to support AD planning and decision making at the regional/county level. The model considered the effect of policy decisions regarding AD configuration (centralized, uniform, or decentralized) and temporal rate of adoption (slow, mid, or aggressive). It also addressed potential implications in the water–energy–food (WEF) nexus through consideration of environmental, social, economic, and spatial parameters.

The three studies presenting DSTs highlighted the need to further develop indicators to assess circularity, sustainability, and associated nexus interlinkages. Further application of DSTs in different geographic contexts can enhance relevant scientific knowledge. Moreover, current DSTs have not built upon one another. Each adopted a different perspective towards AD planning and the consideration of configuration approaches. The development of commonly agreed DSTs could organize and enhance future AD research and planning, while always accounting for contextual characteristics.

4.4. Nexus Approaches

Nexus approaches address interlinkages between resource systems across sectors and scales and aim to identify and manage relevant trade-offs and synergies [75]. In recent years, they have gained increasing scientific interest as tools to support research for sustainable development (e.g., [76,77]). In the AD context, the authors of [31] characterized nexus approaches as valuable tools to address AD's multifunctional character towards integrated resource management. The multisectoral and multiscalar nature of nexus approaches can also be useful in the development of MCDA tools (see Section 4.3).

Only two papers we reviewed addressed urban AD through a nexus approach, both focusing on the WEF nexus. The authors of [73] referred to the WEF nexus in the context of integrating biowaste valorization with UA. However, they only assessed WEF material and energy flows, and did not conduct comprehensive nexus assessment to address nexus interlinkages explicitly (for a description of the concept of 'nexus assessment', see [78]). The study in [64] developed a DST for AD planning, which addressed WEF nexus implications (e.g., fresh water consumed) as part of a sustainability assessment. Their findings showed that each AD strategy could involve several trade-offs, such as between GHG emissions and social acceptance: decentralized scenarios often require short transport distances that minimize associated GHG emissions; however, the proximity of treatment plants to populated areas is associated with low social acceptance. In return, centralized systems can have a lower 'visual impact', since they are located far from populated areas. Nevertheless, larger transport distances can lead to higher GHG emissions, compared to decentralized approaches. These findings highlighted the value of nexus approaches in identifying potential trade-offs and synergies. Nexus approaches can be further applied in the study of decentralized AD. For example, the authors of [64] encouraged further research on nexus metrics in the urban AD context to increase analytical detail, and thus address complexity of measuring nexus interlinkages. Moreover, among the scales addressed through nexus approaches, the urban scale is essential to consider in the context of AD for urban biowaste treatment and embedding AD in the urban environment.

4.5. *UA as Entry Point for AD Implementation*

AD implementation and its embeddedness in the urban environment can be approached through various entry points (see also Section 3.2.1). Among such entry points, several studies have focused on synergies between AD and UA. In recent years, urban agriculture (UA) has been advocated as a potential contributor to urban sustainability. Scientific literature has explored potential UA benefits ranging across various dimensions such as food security, ecosystem services, social cohesion, and others (e.g., [79,80]). In the CE context, UA can contribute to sustainable urban metabolism, with outputs from one process serving as inputs for another [81,82].

To realize potential benefits, UA must be 'upscaled': UA growing practices need to expand in available land areas (ranging from ground-based land plots to building rooftops). While upscaling, UA's resource requirements increase and must be assessed and managed sustainably. However, the authors of [81] highlighted resource requirements of both current and upscaled systems as largely unexplored. The authors of [83] found scientific knowledge insufficient to reach definite conclusions on UA's expected sustainability impact. They also provided an overview of waste valorization pathways to enhance UA's resource efficiency and explore WEF nexus interlinkages. UA can utilize recovery products from urban waste streams (biowaste but also wastewater, waste heat, and CO_2) to cover its resource requirements such as water for irrigation, nutrients, and energy in the forms of heating and electricity (in the case of advanced UA practices such as greenhouses) [73,83]. The potential for synergies between biowaste management and UA has long been mentioned (e.g., [81,84–86]) but few studies have actually assessed aspects of this potential. The study in [80] addressed the potential for nutrient circularity among opportunities and challenges for UA's future development. The authors of [62] conducted a technoeconomic assessment of a pilot AD plant located in a greenhouse in a park in London, UK. The authors reported challenges to balance supply and demand for digestate: in terms of identifying suitable outlets, but also in promoting digestate to consumers; digestate management at the small scale remains highly unregulated, and there is limited scientific evidence on safety to use in UA (in terms of digestate quality and potential toxic effects). Using the same pilot project as [62], the authors of [87] studied the feasibility of digestate use in cities. They also applied actor network theory (ANT) to assess stakeholder views on digestate management. They identified technical feasibility of onsite treatment and its economic viability as main challenges for implementation. The study in [54] assessed digestate use in PUA through short-distance AD systems, and found significant potential for digestate valorization. Case study findings from Lyon and Glasgow supported that biowaste recovery could reduce UA's carbon footprint [73] (as described in Section 4.1). However, even in upscaled UA scenarios, it is unlikely that UA alone can valorize recovery products fully. To achieve integrated resource management, it is important to explore UA further, as well as other urban outlets for biowaste valorization [63].

In summary, the scientific literature supported that upscaling UA can contribute to sustainable urban metabolism and other potential benefits significantly. Upscaling also requires sustainable management of UA's increasing resource requirements, which can be potentially met using urban waste streams. However, the review findings highlighted that few studies have quantified UA's resource demand and the potential for waste valorization to supply them. Current findings showed that it can be challenging to match supply and demand of resources locally. Future research must explore UA along with other urban outlets that can enable sustainable local valorization loops. Nexus approaches can also assist in identifying interlinkages between UA and AD in the urban context.

5. Opportunities and Challenges of Decentralized AD Approaches

The study of definitions, key planning aspects (Section 3), and emerging research themes (Section 4) led to the identification of key opportunities and challenges that can guide future development of decentralized AD, in terms of scientific research and decision making. Table S10 (see Supplementary Material) presents these opportunities and challenges, and also shows the key papers used to identify them.

In summary, all papers addressed the potential of decentralized approaches to develop local circularity, either implicitly or explicitly. To achieve this potential, it is essential to identify urban outlets to support the development of local valorization loops. Local resource supply needs to match the demand and the sustainability impact of local valorization loops must be assessed. UA and PUA have received attention as potential urban outlets: they need to be further explored in combination with other potential urban outlets. Preliminary findings showed that under certain scenarios, decentralized or combined (combinations of centralized and decentralized) configurations have better environmental performance than centralized options. However, such results are highly context-specific. A treatment system that integrates decentralized AD characteristics and is successful for one city may not work for another, and the results are largely dependent on local characteristics. Moreover, there are several remaining challenges related to digestate post-treatment and valorization. Offsite digestate management may increase the system's environmental impact, largely due to transport-related GHG emissions.

The context-specific results highlighted the need to ground AD planning in integrated approaches based on local characteristics. There is need for further case study research in different urban contexts. Even then, results must be extrapolated with caution. Several current themes can support further research on decentralized AD. Notably, DSTs that apply MCDA approaches and consider characteristics of different system configurations can contribute towards integrated AD planning. The inclusion and development of spatial methods can enhance planning accuracy. Nevertheless, context-specific assessments require local data that are often undocumented or unavailable. Further development of indicators to measure circularity, sustainability impact, and associated nexus interlinkages is also needed. Such indicators can be used in DSTs and contribute to integrated AD planning. AD embeddedness in the urban environment has many aspects that remain unexplored, such as product end uses, relevant stakeholders, market structure, and policy measures in different contexts.

6. Conclusions

Practitioners, policy makers, and the scientific community have shown increasing interest in the potential contribution of decentralized approaches in the context of implementing urban biowaste recovery systems. Through a scoping review, this paper provided a comprehensive overview of current scientific knowledge on decentralized AD approaches for urban biowaste recovery systems. The findings showed that there is limited scientific evidence on the impact of decentralized configuration approaches on resource circularity and sustainability in different urban contexts. However, five emerging research themes were identified: (a) spatial analysis; (b) LCA of environmental performance; (c) DSTs and frameworks; (d) nexus approaches; and (e) UA as an entry point for embeddedness in the urban environment. Opportunities and challenges for planning of decentralized AD exist, which mainly concern: (a) digestate management; (b) the potential for local circularity with product valorization in outlets such as UA; and (c) the development and application of DSTs to support integrated planning. Through the opportunities, challenges, and emerging themes addressed, this study can guide future research and decision making towards integrated planning of AD and biowaste recovery systems in cities. The findings highlighted the comprehensive analysis of configuration approaches as an essential component towards integrated planning. Local conditions and context should also be considered in the development of integrated planning to harness the full potential of biowaste recovery systems.

Supplementary Materials: Supplementary material for this manuscript is openly available in the online repository Zenodo at: https://doi.org/10.5281/zenodo.5647379.

Author Contributions: Conceptualization, E.N.; methodology, E.N.; validation, E.N.; investigation, E.N.; data curation, E.N.; writing—original draft preparation, E.N.; writing—review and editing, E.N., D.K. and V.M.; supervision, V.M. All authors have read and agreed to the published version of the manuscript.

Funding: The research was partially funded by the Swedish Research Council FORMAS through project grant number 2017-00266.

Acknowledgments: E.N., D.K. and V.M. greatly appreciate the financial support from FORMAS. We would also like to thank Professor Semida Silveira and colleagues at the division of Energy Systems, KTH, for their valuable feedback/comments at the initial stages of the paper.

Conflicts of Interest: The authors declare no conflict of interest. The funders had no role in the design of the study; in the collection, analyses, or interpretation of data; in the writing of the manuscript; or in the decision to publish the results.

Abbreviations

ABM	Agent-based model
AD	Anaerobic digestion
ANT	Actor network theory
CE	Circular economy
CHP	Combined heat and power
DST	Decision-support tools
EMFA	Extended material flow analysis
GHG	Greenhouse gas
GIS	Geographic information systems
GWP	Global warming potential
IEA	International Energy Agency
LCA	Life cycle assessment
MCDA	Multicriteria decision analysis
MFA	Material flow analysis
MSW	Municipal solid waste
Mtoe	Million tons of oil equivalent
N, P, K	Nitrogen, phosphorus, potassium
PRISMA	Preferred Reporting Items for Systematic Reviews and Meta-Analyses
OFMSW	Organic fraction of municipal solid waste
PUA	Periurban agriculture
STARR	Systems Thinking Approach to Resource Recovery (framework)
WEF	Water–energy–food (nexus)
WoS	Web of Science
WtE	Waste-to-energy
UA	Urban agriculture

References

1. United Nations. The New Urban Agenda—Habitat III. Available online: http://habitat3.org/the-new-urban-agenda/ (accessed on 19 June 2021).
2. Kaza, S.; Yao, L.C.; Bhada-Tata, P.; Van Woerden, F. What a Waste 2.0: A Global Snapshot of Solid Waste Management to 2050. 2018. Available online: https://openknowledge.worldbank.org/handle/10986/30317 (accessed on 19 June 2021).
3. Iacovidou, E.; Millward-Hopkins, J.; Busch, J.; Purnell, P.; Velis, C.A.; Hahladakis, J.N.; Zwirner, O.; Brown, A. A pathway to circular economy: Developing a conceptual framework for complex value assessment of resources recovered from waste. *J. Clean. Prod.* **2017**, *168*, 1279–1288. [CrossRef]
4. Korhonen, J.; Honkasalo, A.; Seppälä, J. Circular Economy: The Concept and its Limitations. *Ecol. Econ.* **2018**, *143*, 37–46. [CrossRef]
5. European Commission. Closing the Loop: An. EU Action Plan for the Circular Economy. 2015. Available online: https://eur-lex.europa.eu/legal-content/EN/TXT/?uri=CELEX%3A52015DC0614 (accessed on 19 June 2021).
6. Köhler, J.; Geels, F.W.; Kern, F.; Markard, J.; Onsongo, E.; Wieczorek, A.; Alkemade, F.; Avelino, F.; Bergek, A.; Boons, F.; et al. An agenda for sustainability transitions research: State of the art and future directions. *Environ. Innov. Soc. Transit.* **2019**, *31*, 1–32. [CrossRef]
7. Merli, R.; Preziosi, M.; Acampora, A. How do scholars approach the circular economy? A systematic literature review. *J. Clean. Prod.* **2018**, *178*, 703–722. [CrossRef]
8. Ingrao, C.; Faccilongo, N.; Di Gioia, L.; Messineo, A. Food waste recovery into energy in a circular economy perspective: A comprehensive review of aspects related to plant operation and environmental assessment. *J. Clean. Prod.* **2018**, *184*, 869–892. [CrossRef]
9. C40 Cities Climate Leadership Group. Good Practice Guide Sustainable Solid Waste Systems. 2015, pp. 1–22. Available online: https://c40-production-images.s3.amazonaws.com/good_practice_briefings/images/9_C40_GPG_SSWS.original.pdf?1456789082 (accessed on 19 June 2021).
10. Finnveden, G.; Johansson, J.; Lind, P.; Moberg, Å. Life cycle assessment of energy from solid waste—Part 1: General methodology and results. *J. Clean. Prod.* **2005**, *13*, 213–229. [CrossRef]
11. Venkata Mohan, S.; Nikhil, G.N.; Chiranjeevi, P.; Nagendranatha Reddy, C.; Rohit, M.V.; Kumar, A.N.; Sarkar, O. Waste biorefinery models towards sustainable circular bioeconomy: Critical review and future perspectives. *Bioresour. Technol.* **2016**, *215*, 2–12. [CrossRef]
12. European Parliament. Directive 2008/98/EC of the European Parliament and of the Council of 19 November 2008 on Waste and Repealing Certain Directives. 2008, pp. 3–30. Available online: http://data.europa.eu/eli/dir/2008/98/oj (accessed on 19 June 2021).
13. Kisser, J.; Wirth, M.; De Gusseme, B.; Van Eekert, M.; Zeeman, G.; Schoenborn, A.; Vinnerås, B.; Finger, D.C.; Kolbl Repinc, S.; Bulc, T.G.; et al. A review of nature-based solutions for resource recovery in cities. *Blue-Green Syst.* **2020**, *2*, 138–172. [CrossRef]
14. Vea, E.B.; Martinez-Sanchez, V.; Thomsen, M. A review of waste management decision support tools and their ability to assess circular biowaste management systems. *Sustainability* **2018**, *10*, 40–60.
15. Directive (EU) 2018/851. Directive (EU) 2018/851 of the European Parliament and of the Council of 30 May 2018 amending Directive 2008/98/EC on Waste. 2018. Available online: http://data.europa.eu/eli/dir/2018/851/oj (accessed on 19 June 2021).
16. European Commission. Review of the 2012 European Bioeconomy Strategy—Publications Office of the EU. 2018. Available online: https://op.europa.eu/en/publication-detail/-/publication/c2f36c72-2e59-11e8-b5fe-01aa75ed71a1/language-en/ (accessed on 19 June 2021).
17. Zeller, V.; Lavigne, C.; D'Ans, P.; Towa, E.; Achten, W.M.J. Assessing the environmental performance for more local and more circular biowaste management options at city-region level. *Sci. Total Environ.* **2020**, *745*, 140690. [CrossRef]
18. Kuznetsova, E.; Cardin, M.A.; Diao, M.; Zhang, S. Integrated decision-support methodology for combined centralized-decentralized waste-to-energy management systems design. *Renew. Sustain. Energy Rev.* **2019**, *103*, 477–500. [CrossRef]
19. Paes, L.A.B.; Bezerra, B.S.; Deus, R.M.; Jugend, D.; Battistelle, R.A.G. Organic solid waste management in a circular economy perspective—A systematic review and SWOT analysis. *J. Clean. Prod.* **2019**, *239*, 118086. [CrossRef]
20. Lohri, C.R.; Diener, S.; Zabaleta, I.; Mertenat, A.; Zurbrügg, C. Treatment technologies for urban solid biowaste to create value products: A review with focus on low- and middle-income settings. *Rev. Environ. Sci. Biotechnol.* **2017**, *16*, 81–130. [CrossRef]
21. Pan, S.Y.; Du, M.A.; Huang, I.T.; Liu, I.H.; Chang, E.E.; Chiang, P.C. Strategies on implementation of waste-to-energy (WTE) supply chain for circular economy system: A review. *J. Clean. Prod.* **2015**, *108*, 409–421. [CrossRef]
22. Escobar, N.; Laibach, N. Sustainability check for bio-based technologies: A review of process-based and life cycle approaches. *Renew. Sustain. Energy Rev.* **2021**, *135*, 110213. [CrossRef]
23. Lytras, G.; Lytras, C.; Mathioudakis, D.; Papadopoulou, K.; Lyberatos, G. Food Waste Valorization Based on Anaerobic Digestion. *Waste Biomass. Valorization* **2021**, *12*, 1677–1697. [CrossRef]
24. Carrere, H.; Antonopoulou, G.; Affes, R.; Passos, F.; Battimelli, A.; Lyberatos, G.; Ferrer, I. Review of feedstock pretreatment strategies for improved anaerobic digestion: From lab-scale research to full-scale application. *Bioresour. Technol.* **2016**, *199*, 386–397. [CrossRef]

25. Fagerström, A.; Al Seadi, T.; Rasi, S.; Briseid, T. *The Role of Anaerobic Digestion and Biogas in the Circular Economy*; Murphy, J.D., Ed.; 2018; Available online: https://www.ieabioenergy.com/wp-content/uploads/2018/08/anaerobic-digestion_web_END.pdf (accessed on 19 June 2021).
26. Poudel, R.C.; Khatiwada, D.; Aryal, P.; Sapkota, M. Large-scale biogas upgrading plants: Future prospective and technical challenges. In *Emerging Technologies and Biological Systems for Biogas Upgrading*; Academic Press: Cambridge, MA, USA, 2021; pp. 467–491. [CrossRef]
27. Stoknes, K.; Scholwin, F.; Krzesiński, W.; Wojciechowska, E.; Jasińska, A. Efficiency of a novel "Food to waste to food" system including anaerobic digestion of food waste and cultivation of vegetables on digestate in a bubble-insulated greenhouse. *Waste Manag.* **2016**, *56*, 466–476. [CrossRef]
28. Lindfors, A.; Feiz, R.; Eklund, M.; Ammenberg, J. Assessing the potential, performance and feasibility of urban solutions: Methodological considerations and learnings from biogas solutions. *Sustainability* **2019**, *11*, 3756. [CrossRef]
29. World Biogas Association. *The Contribution of Anaerobic Digestion and Biogas towards Achieving the UN Sustainable Development Goals*; World Biogas Association: London, UK, 2018; Available online: https://www.worldbiogasassociation.org/wp-content/uploads/2018/12/WBA_SDG_Biogas_Report.pdf (accessed on 19 June 2021).
30. Mao, C.; Feng, Y.; Wang, X.; Ren, G. Review on research achievements of biogas from anaerobic digestion. *Renew. Sustain. Energy Rev.* **2015**, *45*, 540–555. [CrossRef]
31. Hoolohan, C.; Soutar, I.; Suckling, J.; Druckman, A.; Larkin, A.; McLachlan, C. Stepping-up innovations in the water–energy–food nexus: A case study of anaerobic digestion in the UK. *Geogr. J.* **2019**, *185*, 391–405. [CrossRef]
32. Lohri, C.R.; Rodić, L.; Zurbrügg, C. Feasibility assessment tool for urban anaerobic digestion in developing countries. *J. Environ. Manag.* **2013**, *126*, 122–131. [CrossRef] [PubMed]
33. Breitenmoser, L.; Gross, T.; Huesch, R.; Rau, J.; Dhar, H.; Kumar, S.; Hugi, C.; Wintgens, T. Anaerobic digestion of biowastes in India: Opportunities, challenges and research needs. *J. Environ. Manag.* **2019**, *236*, 396–412. [CrossRef]
34. Edwards, J.; Othman, M.; Burn, S. A review of policy drivers and barriers for the use of anaerobic digestion in Europe, the United States and Australia. *Renew. Sustain. Energy Rev.* **2015**, *52*, 815–828. [CrossRef]
35. Nevzorova, T.; Kutcherov, V. Barriers to the wider implementation of biogas as a source of energy: A state-of-the-art review. *Energy Strateg. Rev.* **2019**, *26*, 100414. [CrossRef]
36. International Energy Agency. *Outlook for Biogas and Biomethane*; IEA Publications: Paris, France, 2020; Available online: https://www.iea.org/reports/outlook-for-biogas-and-biomethane-prospects-for-organic-growth (accessed on 19 June 2021).
37. Bortolotti, A.; Kampelmann, S.; De Muynck, S. Decentralised Organic Resource Treatments—Classification and comparison through Extended Material Flow Analysis. *J. Clean. Prod.* **2018**, *183*, 515–526. [CrossRef]
38. Ng, K.S.; Yang, A.; Yakovleva, N. Sustainable waste management through synergistic utilisation of commercial and domestic organic waste for efficient resource recovery and valorisation in the UK. *J. Clean. Prod.* **2019**, *227*, 248–262. [CrossRef]
39. Markard, J.; Raven, R.; Truffer, B. Sustainability transitions: An emerging field of research and its prospects. *Res. Policy* **2012**, *41*, 955–967. [CrossRef]
40. UN Habitat. *Collection of Municipal Solid Waste in Developing Countries*; UN Habitat: Nairobi, Kenya, 2010; Volume 70.
41. Angouria-Tsorochidou, E.; Teigiserova, D.A.; Thomsen, M. Limits to circular bioeconomy in the transition towards decentralized biowaste management systems. *Resour. Conserv. Recycl.* **2021**, *164*, 105207. [CrossRef]
42. Wilson, C.; Grubler, A.; Bento, N.; Healey, S.; De Stercke, S.; Zimm, C. Granular technologies to accelerate decarbonization. *Science* **2020**, *368*, 36–39. [CrossRef]
43. Mangoyana, R.B.; Smith, T.F. Decentralised bioenergy systems: A review of opportunities and threats. *Energy Policy* **2011**, *39*, 1286–1295. [CrossRef]
44. Vasco-Correa, J.; Khanal, S.; Manandhar, A.; Shah, A. Anaerobic digestion for bioenergy production: Global status, environmental and techno-economic implications, and government policies. *Bioresour. Technol.* **2018**, *247*, 1015–1026. [CrossRef] [PubMed]
45. Särkilahti, M.; Kinnunen, V.; Kettunen, R.; Jokinen, A.; Rintala, J. Replacing centralised waste and sanitation infrastructure with local treatment and nutrient recycling: Expert opinions in the context of urban planning. *Technol. Forecast. Soc. Chang.* **2017**, *118*, 195–204. [CrossRef]
46. Seuring, S.; Gold, S. Conducting content-analysis based literature reviews in supply chain management. *Supply Chain Manag.* **2012**, *17*, 544–555. [CrossRef]
47. Munn, Z.; Peters, M.D.J.; Stern, C.; Tufanaru, C.; McArthur, A.; Aromataris, E. Systematic review or scoping review? Guidance for authors when choosing between a systematic or scoping review approach. *BMC Med. Res. Methodol.* **2018**, *18*, 1–7. [CrossRef]
48. Moher, D.; Liberati, A.; Tetzlaff, J.; Altman, D.G.; Altman, D.; Antes, G.; Atkins, D.; Barbour, V.; Barrowman, N.; Berlin, J.A.; et al. Preferred reporting items for systematic reviews and meta-analyses: The PRISMA statement. *PLoS Med.* **2009**, *6*, e1000097-6. [CrossRef] [PubMed]
49. Heyvaert, M.; Hannes, K.; Onghena, P. *Using Mixed Methods Research Synthesis for Literature Reviews*; SAGE Publications, Inc.: Thousand Oaks, CA, USA, 2017; pp. 23–68.
50. Comber, A.; Dickie, J.; Jarvis, C.; Phillips, M.; Tansey, K. Locating bioenergy facilities using a modified GIS-based location-allocation-algorithm: Considering the spatial distribution of resource supply. *Appl. Energy* **2015**, *154*, 309–316. [CrossRef]

51. Popay, J.; Roberts, H.; Sowden, A.; Petticrew, M.; Arai, L.; Rodgers, M.; Britten, N.; Roen, K.; Duffy, S. *Guidance on the Conduct of Narrative Synthesis in Systematic Reviews: A Product from the ESRC Methods Programme*; Lancaster University: Lancaster, UK, 2006; Available online: https://www.lancaster.ac.uk/media/lancaster-university/content-assets/documents/fhm/dhr/chir/NSsynthesisguidanceVersion1-April2006.pdf (accessed on 19 June 2021).
52. Snilstveit, B.; Oliver, S.; Vojtkova, M. Narrative approaches to systematic review and synthesis of evidence for international development policy and practice. *J. Dev. Eff.* **2012**, *4*, 409–429. [CrossRef]
53. Braun, V.; Clarke, V. Using thematic analysis in psychology. *Qual. Res. Psychol.* **2006**, *3*, 77–101. [CrossRef]
54. Thiriet, P.; Bioteau, T.; Tremier, A. Optimization method to construct micro-anaerobic digesters networks for decentralized biowaste treatment in urban and peri-urban areas. *J. Clean. Prod.* **2020**, *243*, 118478. [CrossRef]
55. Curry, N.; Pillay, P. Biogas prediction and design of a food waste to energy system for the urban environment. *Renew. Energy* **2012**, *41*, 200–209. [CrossRef]
56. Righi, S.; Oliviero, L.; Pedrini, M.; Buscaroli, A.; Della Casa, C. Life Cycle Assessment of management systems for sewage sludge and food waste: Centralized and decentralized approaches. *J. Clean. Prod.* **2013**, *44*, 8–17. [CrossRef]
57. Guilayn, F.; Rouez, M.; Crest, M.; Patureau, D.; Jimenez, J. Valorization of digestates from urban or centralized biogas plants: A critical review. *Rev. Environ. Sci. Biotechnol.* **2020**, *19*, 419–462. [CrossRef]
58. Zeller, V.; Towa, E.; Degrez, M.; Achten, W.M.J. Urban waste flows and their potential for a circular economy model at city-region level. *Waste Manag.* **2019**, *83*, 83–94. [CrossRef] [PubMed]
59. Bernstad, A.; la Cour Jansen, J. Separate collection of household food waste for anaerobic degradation—Comparison of different techniques from a systems perspective. *Waste Manag.* **2012**, *32*, 806–815. [CrossRef] [PubMed]
60. Sisto, R.; Sica, E.; Lombardi, M.; Prosperi, M. Organic fraction of municipal solid waste valorisation in southern Italy: The stakeholders' contribution to a long-term strategy de fi nition. *J. Clean. Prod.* **2017**, *168*, 302–310. [CrossRef]
61. Di Matteo, U.; Nastasi, B.; Albo, A.; Astiaso Garcia, D. Energy contribution of OFMSW (Organic Fraction of Municipal Solid Waste) to energy-environmental sustainability in urban areas at small scale. *Energies* **2017**, *10*, 229. [CrossRef]
62. Walker, M.; Theaker, H.; Yaman, R.; Poggio, D.; Nimmo, W.; Bywater, A.; Blanch, G.; Pourkashanian, M. Assessment of micro-scale anaerobic digestion for management of urban organic waste: A case study in London, UK. *Waste Manag.* **2017**, *61*, 258–268. [CrossRef] [PubMed]
63. Weidner, T.; Graça, J.; Machado, T.; Yang, A. Comparison of local and centralized biowaste management strategies—A spatially-sensitive approach for the region of Porto. *Waste Manag.* **2020**, *118*, 552–562. [CrossRef]
64. Falconer, R.E.; Haltas, I.; Varga, L.; Forbes, P.J.; Abdel-Aal, M.; Panayotov, N. Anaerobic Digestion of food waste: Eliciting sustainable water-energy-food nexus practices with Agent Based Modelling and visual analytics. *J. Clean. Prod.* **2020**, *255*, 120060. [CrossRef]
65. Rajendran, K.; Aslanzadeh, S.; Taherzadeh, M.J. Household Biogas Digesters—A Review. *Energies* **2012**, *5*, 2911–2942. [CrossRef]
66. Garfí, M.; Martí-Herrero, J.; Garwood, A.; Ferrer, I. Household anaerobic digesters for biogas production in Latin America: A review. *Renew. Sustain. Energy Rev.* **2016**, *60*, 599–614. [CrossRef]
67. Chodkowska-Miszczuk, J.; Martinat, S.; Cowell, R. Community tensions, participation, and local development: Factors affecting the spatial embeddedness of anaerobic digestion in Poland and the Czech Republic. *Energy Res. Soc. Sci.* **2019**, *55*, 134–145. [CrossRef]
68. Bautista Angeli, J.R.; LeFloc'h, T.; Lakel, A.; Lacarrière, B.; Andres, Y. Anaerobic digestion of urban wastes: Integration and benefits of a small-scale system. *Environ. Technol.* **2021**, 1–12. [CrossRef]
69. Iacovidou, E.; Velis, C.A.; Purnell, P.; Zwirner, O.; Brown, A.; Hahladakis, J.; Millward-Hopkins, J.; Williams, P.T. Metrics for optimising the multi-dimensional value of resources recovered from waste in a circular economy: A critical review. *J. Clean. Prod.* **2017**, *166*, 910–938. [CrossRef]
70. Bautista Angeli, J.R.; Morales, A.; LeFloc'h, T.; Lakel, A.; Andres, Y. Anaerobic digestion and integration at urban scale: Feedback and comparative case study. *Energy. Sustain. Soc.* **2018**, *8*, 29. [CrossRef]
71. Gailing, L.; Bues, A.; Kern, K.; Röhring, A. Socio-spatial dimensions in energy transitions: Applying the TPSN framework to case studies in Germany. *Environ. Plan. A* **2019**, *52*, 1–19. [CrossRef]
72. Van der Wiel, B.Z.; Weijma, J.; van Middelaar, C.E.; Kleinke, M.; Buisman, C.J.N.; Wichern, F. Restoring nutrient circularity: A review of nutrient stock and flow analyses of local agro-food-waste systems. *Resour. Conserv. Recycl.* **2020**, *160*, 104901. [CrossRef]
73. Weidner, T.; Yang, A. The potential of urban agriculture in combination with organic waste valorization: Assessment of resource flows and emissions for two european cities. *J. Clean. Prod.* **2020**, *244*, 118490. [CrossRef]
74. Tian, H.; Wang, X.; Lim, E.Y.; Lee, J.T.E.; Ee, A.W.L.; Zhang, J.; Tong, Y.W. Life cycle assessment of food waste to energy and resources: Centralized and decentralized anaerobic digestion with different downstream biogas utilization. *Renew. Sustain. Energy Rev.* **2021**, *150*, 111489. [CrossRef]
75. Hoff, H. *Understanding the Nexus. Background Paper for the Bonn2011 Conference: The Water, Energy and Food Security Nexus*; Stockholm Environment Institute: Stockholm, Sweden, 2011; pp. 1–52. Available online: https://mediamanager.sei.org/documents/Publications/SEI-Paper-Hoff-UnderstandingTheNexus-2011.pdf (accessed on 19 June 2021).
76. Liu, J.; Hull, V.; Godfray, H.C.J.; Tilman, D.; Gleick, P.; Hoff, H.; Pahl-Wostl, C.; Xu, Z.; Chung, M.G.; Sun, J.; et al. Nexus approaches to global sustainable development. *Nat. Sustain.* **2018**, *1*, 466–476. [CrossRef]

77. Bleischwitz, R.; Spataru, C.; VanDeveer, S.D.; Obersteiner, M.; van der Voet, E.; Johnson, C.; Andrews-Speed, P.; Boersma, T.; Hoff, H.; van Vuuren, D.P. Resource nexus perspectives towards the United Nations Sustainable Development Goals. *Nat. Sustain.* **2018**, *1*, 737–743. [CrossRef]
78. Ramos, E.P.; Howells, M.; Sridharan, V.; Engström, R.E.; Taliotis, C.; Mentis, D.; Gardumi, F.; de Strasser, L.; Pappis, I.; Balderrama, G.P.; et al. The climate, land, energy, and water systems (CLEWs) framework: A retrospective of activities and advances to 2019. *Environ. Res. Lett.* **2021**, *16*. [CrossRef]
79. Artmann, M.; Sartison, K. The role of urban agriculture as a nature-based solution: A review for developing a systemic assessment framework. *Sustainability* **2018**, *10*, 1937. [CrossRef]
80. Weidner, T.; Yang, A.; Hamm, M.W. Consolidating the current knowledge on urban agriculture in productive urban food systems: Learnings, gaps and outlook. *J. Clean. Prod.* **2019**, *209*, 1637–1655. [CrossRef]
81. Goldstein, B.; Hauschild, M.; Fernández, J.; Birkved, M. Urban versus conventional agriculture, taxonomy of resource profiles: A review. *Agron. Sustain. Dev.* **2016**, *36*, 1–19. [CrossRef]
82. Wielemaker, R.C.; Weijma, J.; Zeeman, G. Harvest to harvest: Recovering nutrients with New Sanitation systems for reuse in Urban Agriculture. *Resour. Conserv. Recycl.* **2018**, *128*, 426–437. [CrossRef]
83. Mohareb, E.; Heller, M.; Novak, P.; Goldstein, B.; Fonoll, X.; Raskin, L. Considerations for reducing food system energy demand while scaling up urban agriculture. *Environ. Res. Lett.* **2017**, *12*, 125004. [CrossRef]
84. Mougeot, L.J.A. Urban Agriculture: Definition, Presence, Potentials and Risks, and Policy Challenges. 2000. Available online: https://idl-bnc-idrc.dspacedirect.org/bitstream/handle/10625/26429/117785.pdf?sequence=12.%20 (accessed on 19 June 2021).
85. Van Timmeren, A. Waste equals energy: Decentralised anaerobic waste treatment and local reuse of return flows. *WIT Trans. Ecol. Environ.* **2007**, *105*, 267–276. [CrossRef]
86. Grewal, S.S.; Grewal, P.S. Can cities become self-reliant in food? *Cities* **2012**, *29*, 1–11. [CrossRef]
87. Fuldauer, L.I.; Parker, B.M.; Yaman, R.; Borrion, A. Managing anaerobic digestate from food waste in the urban environment: Evaluating the feasibility from an interdisciplinary perspective. *J. Clean. Prod.* **2018**, *185*, 929–940. [CrossRef]

MDPI
St. Alban-Anlage 66
4052 Basel
Switzerland
Tel. +41 61 683 77 34
Fax +41 61 302 89 18
www.mdpi.com

Sustainability Editorial Office
E-mail: sustainability@mdpi.com
www.mdpi.com/journal/sustainability

www.ingramcontent.com/pod-product-compliance
Ingram Content Group UK Ltd.
Pitfield, Milton Keynes, MK11 3LW, UK
UKHW060124010826
14089UKWH00006B/143

9 783036 54343